GENERAL DYNAMICS
F-16 FIGHTING FALCON

1978 onwards (all marks)

COVER CUTAWAY:

F-16C Block 40. *(Mike Badrocke)*

First published in February 2014

A catalogue record for this book is available from the British Library

ISBN 978 0 85733 398 8

Library of Congress control no. 2013954587

Published by Haynes Publishing,
Sparkford, Yeovil,
Somerset BA22 7JJ, UK.
Tel: 01963 442030 Fax: 01963 440001
Int. tel: +44 1963 442030 Int. fax: +44 1963 440001
E-mail: sales@haynes.co.uk
Website: www.haynes.co.uk

Haynes North America Inc.,
861 Lawrence Drive, Newbury Park,
California 91320, USA.

Printed in the USA by Odcombe Press LP,
1299 Bridgestone Parkway, La Vergne, TN 37086.

Acknowledgements

The author would like to thank the following for their assistance in the creation of this book: Major Sean 'Lobo' Canfield; Lieutenant Josh Smith; and Bjorn Claes, editor of the web's premier F-16 resource, www.f-16.net.

GENERAL DYNAMICS
F-16 FIGHTING FALCON

1978 onwards (all marks)

Owners' Workshop Manual

An insight into operating, maintaining and flying
the USAF all-weather, multi-role jet fighter

Steve Davies

Contents

(USAF)

Chapter One

The Viper story

●

Conceived and marketed as the ultimate point defence fighter, more than 4,500 Lockheed Martin (formerly General Dynamics) F-16 Fighting Falcons had been manufactured by 2013, making the type one of the most ubiquitous fighters in operational use today. With 2,231 F-16s delivered to the United States Air Force (USAF) between December 1976 and January 2009, it is also the most prolific fighter in US service.

OPPOSITE The YF-16 in flight. Very small by the standards of the day, the 'electric jet' was initially conceived as a lightweight fighter that would dominate an adversary during a close-in fight. *(USAF)*

The F-16 was originally designed as a lightweight, short-range point defence fighter. However, in the intervening three and a half decades it has evolved into a truly multi-role platform, as much at home with ground attack as it is with the more glamorous business of downing other fighters.

Lessons learned

The F-16 design came about as a result of lessons learned following the experiences of air combat by the USAF, US Navy and US Marine Corps during the Vietnam War of the 1960s and early 1970s. That war had aptly demonstrated that, whilst technological prowess was important, America had somewhat overestimated how it could be brought to bear in actual combat conditions.

The war also served as a stark reminder that there was no substitute for getting the basics right. Large, complex and heavy US-made fighters had fared poorly against small, light, simple and highly manoeuvrable Russian-designed MiGs.

On paper, America's advantage of long-range missiles that could reach out and kill the enemy at extended distances was impressive, but in reality this benefit was negated by an inability to confidentially identify the target as friend of foe until it was well within visual range. Once there – at the so-called 'merge' – long-range missiles designed to down lumbering bombers offered little advantage.

Poor air combat training had also hamstrung American pilots when they tackled North Vietnamese Air Force MiGs and began the classic dogfight. At this juncture, crews flying America's premier fighter – the McDonnell F-4 Phantom – found themselves at a significant disadvantage, since early versions of the Phantom had no internal cannon or Gatling gun with which to kill the enemy, and their short-range air-to-air missiles offered a 'probability of

LEFT The Lightweight Fighter Competition sought to develop a new aircraft that would replace a number of ageing airframes: the F-4 Phantom, A-7E Corsair II and the F-104 Starfighter among them. *(USAF)*

kill' that was so low as to instil little confidence. The lesson was simple: in future there would be no excuse for building any fighter without a gun.

In the United States, one man began applying his own personal experience of flying fighters during the Korean War (1950–53) and, with the warning signs already evident in the skies over North Vietnam, he turned to a mathematical model to provide a quantitative solution to how best to fight the little MiGs. Colonel John Boyd, a controversial figure at the time because of his unconventional thinking, developed the 'Energy-Manoeuvrability' theory (EM) with the assistance of mathematician Thomas Christie.

The EM graph provides a method of expressing an aircraft's performance as a total of kinetic and potential manoeuvrability. In plain English, it shows turn rate plotted against Mach number, allowing a pilot to see the optimum manoeuvring performance of his aircraft. When overlaid on to the EM graph of an adversary aircraft, it graphically illustrates where the strengths and weaknesses lie for each.

Crucially, in 1969 the EM concept gave Boyd (and a cluster of like-minded fighter pilots assigned to the Pentagon) the quantitative basis from which to set out the requirement for a new lightweight fighter (LWF) that would boast an impressive thrust-to-weight ratio. The LWF would address the shortcomings of America's current inventory of fighters, although principally the F-4 Phantom. And so it was that Boyd and his Pentagon cohorts (often collectively referred to as 'the Fighter Mafia') were able to get Department of Defense approval to commence concept studies for a new species of US fighter aircraft. The two defence contractors chosen to undertake the theoretical studies were General Dynamics and Northrop.

By January 1972, an RFP (request for proposal) had been drafted to formally invite a range of defence contractors to submit designs for a 20,000lb fighter, with an excellent turn rate and acceleration, for a flyaway cost of $3 million per unit. The design would need to be optimised for manoeuvring at speeds between Mach 0.6 and 1.6 and at altitudes of 30,000 to 40,000ft, since it was in this portion of the flight envelope that most future air combat was expected to occur. Successful proposals would be entered into the LWF competition.

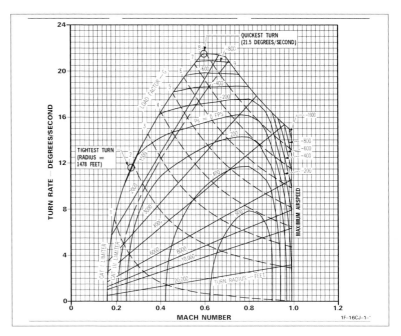

It is worth noting that the LWF concept was itself divisive. It coexisted with the Air Force's F-X programme, which sought a much heavier, all-weather, day-and-night, long-range air superiority fighter with a big radar, twin engines and long-range missiles. (Many sources state, incorrectly, that F-X referred to 'Fighter Experimental', but it actually meant 'Fighter Unknown'.) The F-X programme would eventually spawn the McDonnell Douglas F-15 Eagle (arguably the most successful air superiority fighter of all time), but at the time F-X proponents naturally saw the LWF as a threat to their budgetary allocation.

Importantly at this stage though, there was no commitment on the part of the US Air Force to actually purchase the winning LWF design. The competition was intended to produce technology demonstrators, not operational fighters.

Lightweight Fighter Competition

Although five companies responded to the RFP, it was no surprise that the two design proposals chosen for the LWF competition were General Dynamics' (GD) Model 401 and Northrop's P-600. GD received $38 million to manufacture its Model 401 design (designated YF-16 by the Air Force), while Northrop received $40 million to build its P-600 (YF-17).

ABOVE Boyd and Christie devised the energy manoeuvrability diagram, allowing easy understanding of performance strengths and weaknesses at a glance. Here, the F-16CM EM diagram shows turn rate plotted against Mach number. *(USAF)*

ABOVE The YF-16 and YF-17 in flight during the competition. The YF-17 would go on to become the F/A-18 Hornet. *(USAF)*

BELOW Workmen inspect the fuselage of a YF-16 at the Rome Air Development Center's Newport site. The array of flags along the aircraft's nose point to the multinational interest in the design. *(USAF)*

The Model 401 had evolved from numerous studies that had evaluated an array of configurations on wing design and airframe structures. In addition, General Dynamics had studied the pros and cons of manoeuvring devices, the number of engines and the design of control surfaces, as well as the location and number of vertical stabilisers (tail fins).

In time the company's aerodynamicists had settled on two potential configurations: one with a traditional wing, fuselage and empennage design; the other utilising a blended-wing body with vertical and horizontal stabilisers attached to booms. From these, the most appropriate design features were blended to create the Model 401 design.

While the two companies' prototypes were being prepared for their maiden flights in early 1974, the Fighter Mafia overcame

internal resistance from advocates of the F-15 programme: they proposed that in the future the Air Force could buy a complementary mix of the two. Colloquially referred to as a 'high-low' mix (high-cost F-15, low-cost LWF), the concept was well received. The mix facilitated the creation of a suitably robust force structure with which to fight future wars.

In December 1973 the YF-16 was formally rolled out at the Air Force Flight Test Center, Edwards AFB, California. Although General Dynamics' design first flew accidentally during a high-speed taxi test on 20 January 1974 (it was airborne for six minutes), its official maiden flight took place at Edwards on 2 February. Lasting 90 minutes, this 'first' flight was followed on 6 February by its first supersonic flight. By early May the second YF-16 had also made its first flight. Northrop's two YF-17 prototypes flew for the first time in June and August of the same year respectively.

International interest

The first flights coincided with interest in the LWF from abroad. Several NATO (North Atlantic Treaty Organisation) member states – the Netherlands, Denmark, Belgium and Norway – that were also operating ageing American-built fighter bombers (most commonly the Lockheed F-104 Starfighter) joined together to form the Multinational Fighter Program Group (MFPG). The MFPG would be a collective through which these countries could consider the purchase of whichever design triumphed in the LWF competition.

This international interest, combined with the view from the US Congress that military acquisition programmes needed to be more efficient and more cohesive (a view that culminated in forcing the US Navy and Air Force to work together to acquire common solutions), resulted in the LWF competition being melded into the new Air Combat Fighter (ACF) competition. This move was of singular importance: it took the LWF out of the arena of a technology demonstrator and into that of an eventual operational fighter for which large orders would be placed. The stakes had been raised.

Pleasingly for the Air Force, the US Secretary of Defense also confirmed that the ACF winner

RIGHT AND CENTRE Two views showing the future of the US Air Force's 'high-low' mix of fighters. The LWF competition was initially seen to pose a threat to what would become the F-15 Eagle, but in fact the two turned out to complement one another well. As can be seen, the twin-engined, twin-tailed Eagle is far larger than the F-16. There are strengths and weaknesses to each: the larger Eagle carries a bigger radar with which to see further, but it is a large visual target visible as far away as ten miles. The F-16 lacks the capability to detect fighter-size targets at the same sort of ranges as the Eagle (about 60 miles), but its small visual signature enables it to get to the merge unseen on a regular basis. *(USAF)*

would indeed be bought *in addition to* the F-15, thus confirming the Fighter Mafia's earlier high-low mix proposals. However, the announcement also applied commercial pressure on GD and Northrop – it opened the competition to international contractors who wished to submit their own aircraft designs for consideration.

Despite submissions by a number of international aerospace companies, the flight test programme eventually revealed that the YF-16 was the most manoeuvrable of all the designs, but especially so when compared to France's Dassault-Breguet Mirage F-1M-53 and Sweden's Saab JAS-37E.

The inclusion of the JAS-37E – the export version of the Saab JAS-37 Viggen – was noteworthy. The Viggen made use of canards to control an aerodynamic phenomenon that occurred with highly-swept wings and had the desirable effect of creating additional lift in certain conditions: vortex lift.

Highly-swept wings generate powerful vortices as angle of attack (AoA – the angle of the wing as it cuts through relative airflow) increases, stabilising the aircraft in longitudinal

RIGHT The YF-16 cockpit closely resembled what would follow in production variants. The major difference was the replacement of flight instruments on the centre pedestal and left console with a radar scope and stores panel respectively. *(USAF)*

and directional axes as speed decays and pitch increases. This was an area of aerodynamics that GD had spent considerable time and resources investigating.

The trailing-edge vortex that flowed from the Viggen's canards improved the lift of the main wings behind and below them. Moreover, the two canards (one either side of, and aft of, each air intake) also created a leading-edge vortex that was strong enough to improve lift over the main wing as AoA increased. GD had studied the effects of such canards, but chose a different solution to harness the 'power' of vortex lift. It opted to sharpen the leading edge of the YF-16 'forebody', thereby increasing the strength of the vortices that separated and flowed over the wing as angle of attack increased. This 'separated flow' provided additional lift over the main wing and also stabilised the airflow across the outer surfaces of the wings. Additional stability came courtesy of a strake that led to the vertical stabiliser.

Earlier studies by GD had shown that the addition of leading-edge and trailing-edge flaps – surfaces that effectively changed the shape of the wing – complemented the wing-body strake by providing increased directional stability at high angles of attack. The leading-edge flaps were programmed to deploy automatically, increasing lift by as much as 18% and reducing drag by as much as 22% when performing a sustained turn at Mach 0.9 at 30,000ft. They also reduced the drag at maximum lift by nearly 70%. These additions further increased the stability-enhancing qualities of the vertical stabiliser and were therefore included in the YF-16 design.

Relaxed static stability and fly-by-wire

In the course of the competition, the YF-16 had flown 330 sorties for 417 hours of flight time. In many of these sorties it was evaluated at high angles of attack, where separated airflow from the wing and wing-body caused sudden decreases in stability and controllability. Such issues would do little to make the design one that provided the pilot with carefree handling across the flight envelope, so these handling characteristics had to be ironed out.

High angle of attack considerations influenced another of the distinguishing features of the YF-16: the configuration of the engine air inlet. Mounted behind and below the cockpit, it was designed to allow air to enter the intake relatively undisturbed.

GD had intentionally created the YF-16 to exhibit 'relaxed static stability': its centre of lift is located forward of its centre of gravity (whereas it is traditionally located aft of it). Relaxed static stability allowed the design to achieve an initial pitch rate of 5g per second, which was about two and a half times that attainable by the F-4C Phantom. However, it also meant that the aircraft required a computer in order to create 'artificial' stability by very rapidly adjusting the flight control surfaces. This was essential to

BELOW (Left) Painted to evaluate the effectiveness of different camouflage patterns, the YF-16 shows off its most streamlined configuration of two AIM-9 Sidewinder missiles on the wingtips. (Right) Later paint schemes saw the introduction of a wraparound grey scheme and a black radome. *(USAF)*

keep the aircraft from departing controlled flight – a task that requires such rapid and precise corrections that it is beyond the scope of human ability.

This flight control computer is often termed a 'fly-by-wire' control system, and had in fact been developed prior to the implementation of the relaxed static stability design of the YF-16. Indeed, the existence of the advanced flight control computer actually permitted the dynamics of the Model 401 design to be changed.

Whereas traditional flight control systems utilise pulleys, rods and cables connected to the pilot's controls, the computerised flight control system took roll, pitch, and yaw commands from the pilot and then decided how much control surface deflection was required in order to achieve the desired result. Of course, the same computer simultaneously applied the required inputs to keep the relaxed design actually flying. In short, it gave the pilot very precise control with very little lag.

GD's system is more accurately described as a 'quad-redundant (fail-operative, fail-safe), high-authority, command-and-stability augmentation system', as Harry J. Hillaker, former F-16 deputy programme director, wrote in a paper for the National Academy of Engineering in 2004. 'The system consists of a series of sensors (accelerometers, rate gyros, air data converter), computers, selectors, transducers, and inverters that collectively generate the pitch, roll, and yaw rates that are transmitted as electronic signals to the five triplex electrohydraulic, servoactuators that control the flaperons (roll and flaps), elevons (pitch and roll), and rudder.

'The weight saving resulting from the absence of cables, linkages, bell cranks, and the ratio changer was translated into redundancy,' Hillaker added. 'The redundancy level and the freedom of routing afforded by wire harnesses improved the reliability and increased the operational survivability of the airplane and contributed to its compactness and small size. With a conventional flight control system, we would have had to route the cables externally, as on the old Ford Tri-motor, to maintain the size or else make the airplane bigger.'

Even though the fly-by-wire system artificially limited the aircraft's angle of attack,

LEFT **Pressure suits were worn during evaluation of the YF-16's high-altitude performance envelope. The heavily reclined seat is obvious in this image.** *(USAF)*

the competition – and subsequent explorative test flights – revealed that it was still possible to enter a deep stall. This happened when the angle of attack increased so far that the YF-16 would pitch up, but lacked sufficient aerodynamic control to get the nose back down again. This would be addressed only later by increasing the size of the elevons about 25% and by providing the pilot with a manual pitch override (MPO) switch that cut out the computer and gave him direct control.

On 13 January 1975 the Secretary of the Air Force announced that the YF-16 had won the competition. In addition to unadulterated manoeuvrability (particularly at supersonic speeds), the YF-16 out-accelerated the YF-17, offered lower operating costs and had better range. It also offered better commonality with the F-15 because it was powered by the Pratt & Whitney (PW) F100 engine.

Less than a month later, in February, the MFPG of NATO countries was offered the F-16 at a turnkey cost of $5.16 million per aircraft.

Orders

The initial order for the F-16 would be for 650 airframes, although there was an acknowledgement that the figure might rise to as many as 1,400. There would eventually be good news for the YF-17 team too: in 1975 it was selected as the Naval ACF competition winner, eventually entering US Navy service (having undergone significant refinement) as the F/A-18 Hornet.

By June 1975 the European nations had placed an initial order for 348 F-16s (116 for Belgium, 58 for Denmark, 102 for the

ABOVE The two-seat F-16 required the cockpit section to be redesigned. At left, Air Force officers evaluate the configuration. Visibility from the rear cockpit is not exceptional, but it suffices for instructional purposes. The pre-production F-16B differed little from the production variant (right). The second crewmember comes at no expense to the F-16's combat capability, although it does reduce the capacity of a fuselage fuel tank. *(USAF)*

Netherlands and 72 for Norway). Along with approval for a technology transfer for the F100 engine, the United States had sweetened the deal by allowing these export nations to licence-build the aircraft. Final assembly of these airframes would take place in Belgium (SABCA's Gosselies plant), the Netherlands (Fokker's plant at Schiphol) and General Dynamic's Fort Worth plant, Texas. The components would come from a network of contractors: the aft fuselage from SONACA, Belgium; fuselage centre sections, leading- and trailing-edge flaps and flaperons from Fokker; the wing structure and assembly of the wing itself from SABCA; the vertical stabiliser and weapons pylons from Per Udsen, Denmark; the undercarriage from DAF, the Netherlands; the wheels from Raufoss, Norway; the F100 engines from Belgian Fabrique National (FN); and so on.

General Dynamics built eight full-scale development (FSD) F-16A/B airframes for flight-testing. The first FSD F-16A flew at Fort Worth, Texas, on 8 December 1976, while the first two-seat FSD F-16B flew on 8 August 1977. Each was equipped with a Westinghouse AN/APG-66 pulse Doppler radar. For the two-seat B-model, the enlarged canopy and elongated forward fuselage added no additional aerodynamic drag,

although the space required for the second pilot resulted in a 1,500lb reduction of fuel capacity.

The flight-test programme saw the first weight gain for the F-16, and this was to become an ongoing fact of life as each iteration of the aircraft was developed. Of course, this resulted in its thrust-to-weight ratio decreasing, which was contrary to the original vision of the Fighter Mafia.

Compared to the YF-16, the F-16A was lengthened by 10ft. Other significant changes included increasing the wing area from 280ft² to 300ft², decreasing the height of the vertical stabiliser and enlarging the ventral fins, adding two additional stores stations to the wings, lengthening the nose to accommodate the radar, and replacing the nose-wheel double doors with a single door. This all further increased the fighter's weight by 25%.

Whilst such gains were arguably unavoidable during development, the Air Force's increasing focus on developing the F-16's air-to-ground capability meant the aircraft was only ever going to get heavier. The Air Force recognised that it could not commission and purchase a dedicated point defence fighter when the United States would more than likely find itself fighting its wars in far-off places. Far-off places meant complex and costly logistics, and that meant that every bit of 'iron on the ramp' had to provide as much capability to theatre commanders as possible.

Seeking more capability in the ground attack arena, the Air Force took a sideways glance at the F-15 Eagle, but the big fighter only offered very basic air-to-ground capabilities – the F-15 programme office had staunchly resisted all

attempts to add any significant bomb-dropping capabilities, a position nicely summarised by their coining of the phrase 'Not a pound [of weight] for air-to-ground!' In any case, the Eagle was seen as being too expensive and too important to be assigned to close air support, battlefield interdiction and other ground-attack roles.

As a result it fell to the F-16 to take up the slack and fill the air-to-ground mission shortfall. That meant additional avionics equipment, or, measured another way, even more weight.

Production for the US Air Force and European nations

In 1977 the Air Force confirmed plans to purchase 783 additional F-16s, having satisfied itself that what had once been a lightweight fighter could now operate as a multi-role fighter of some potency.

In Belgium, the SABCA production line opened in February 1978, followed two months later by Fokker's production line in the Netherlands. Large-scale production for USAF F-16s began that same year, and the first production F-16A flew at Fort Worth in August. Meanwhile, the air forces of the European nations were now known collectively as the European Participating Air Forces (EPAF).

Into service

Full technical details of each variant and 'Block' are provided in Chapter 2, *Variant summaries and Blocks*.

ABOVE Air refuelling formed an important part of the F-16's evaluation. The air refuelling receptacle is located directly behind the cockpit – a far more convenient location than, say, directly in front of the forward canopy as seen on the A-10 Thunderbolt II. *(USAF)*

BELOW Photographed in 1985, a pre-production F-16A is mounted on a test stand at the Newport Test Site of the Rome Air Development Center. Fitted with AIM-9 Sidewinder missiles, Mk20 Rockeye cluster bombs, a centreline fuel tank and an AN/ALQ-119 electronic countermeasures pod on an outside wing pylon, the aircraft's radar cross-section is being measured. Throughout the F-16's operational life, efforts have been made to reduce its visual, IR and radar signature. *(USAF)*

ABOVE The F-16A/B entered service with the 388th TFW at Hill AFB, Utah. Here, a single-seater prepares for taxi. Note the black radome cover – an obvious external sign that this is a very early F-16A. The black radome would later be changed to grey. *(USAF)*

BELOW Another black-radomed F-16A. Note the two-tone grey scheme adopted for the Viper once it became operational. *(USAF)*

RIGHT Post-refuelling, an F-16A backs off from the tanker with refuelling door still open and the fuselage wet from excess fuel that has vented from the refuelling 'boom' following disconnect. *(USAF)*

F-16A/B: Block 1–Block 15

The first production F-16A was accepted by Tactical Air Command on 17 August 1978, and the first operational unit to operate the type was the 388th Tactical Fighter Wing, Hill Air Force Base, Utah, in January 1979. Hill's 4th Tactical Fighter Squadron became the first to achieve Initial Operational Capability (IOC) on 12 November 1980, meaning that it could, if required, take the type into combat.

The Air Force christened the F-16 the 'Fighting Falcon', but the name instantly met with derision from those assigned to fly it. This, of course was a source of great hilarity to those flying the F-4, F-15 and other fighters. As a result the much more menacing moniker 'Viper' was chosen by the F-16 community as an unofficial alternative. Predictably, other fighter pilots still take delight in referring to it as the Fighting Falcon!

In Europe, the first F-16A delivered from the United States arrived at SABCA on 9 June 1978 for use in assembly tests at the SABCA plant. The first flight of a European-built example took place in December. A little under six months later, Fokker's first F-16 made its maiden flight in May 1979. Deliveries to the EPAF nations began with deliveries to the Belgian Air Force in January 1979. The Royal Dutch Air Force received its first deliveries in June that year, with the Royal Danish Air Force taking deliveries from February 1980 and the Royal Norwegian Air Force in April 1980.

The F-16A/B remained in production for the USAF until 1985, at which point production of the Block 25 F-16C/Ds was already under way (the first examples having already been assembled in early summer 1984).

In October 1986 the USAF announced the conversion of 270 (later changed to 241) F-16A/B Block 15 Fighting Falcons to Air Defense Fighter (ADF) standard. With the Cold War showing no immediate signs of

LEFT F-16A Block 10B, serial 79-0386, was delivered to the Hill AFB in March 1981. Seen here at an RAF station participating in a tactical bombing competition in May 1981, it was lost in a fatal accident in January 1983. Note the smaller radome of the early F-16A/B. *(USAF)*

abating, 14 ANG (Air National Guard) squadrons were slated to receive F-16 ADFs to defend North America.

The F-16 ADF was developed and tested at Edwards AFB, California, in 1990, followed by operational test and evaluation by the 57th Fighter Weapons Wing at Nellis AFB, Nevada. The first service aircraft were assigned to the 114th Fighter Squadron, Oregon ANG, at Kingsley Field, Klamath Falls. Conversion of all 241 airframes (217 A models and 24 B models) was undertaken at the Ogden Air Logistics Center (OALC), Hill AFB, Utah, and had been completed by mid-1992.

AFE – Alternative Fighter Engine

Coincident with the arrival of Block 25, the USAF began the Alternative Fighter Engine (AFE) programme, splitting engine orders between Pratt & Whitney and General Electric. The programme sought to keep prices low by encouraging competition between the two manufacturers. It also created a second source of engines that could be steadily supplied to the logistical supply chain.

The two USAF candidates were the General Electric F101 (now redesignated F110) and a revised F100-PW-100, known as the F100-PW-220. For the latter, Pratt & Whitney would eventually offer an upgrade in kit form for the PW-100, allowing engine maintainers to upgrade the motor at Wing level. The resulting engine is designated the PW-220E.

In February 1984 the USAF announced that General Electric had been awarded 75% of the total engine contracts for the FY 1985 run of F-16 fighters, while the remaining FY 1985 F-16s would use the upgraded Pratt & Whitney F100. The F110 was to be phased into the General Dynamics production line as soon as production engines became available, but it was agreed that individual USAF F-16 units should

ABOVE Pilots who fly the 'Viper' cannot help but be impressed. Manoeuvrable, fast and highly automated, it is a fighter pilot's dream. To those who do not fly it, much fun can be had ridiculing the rather awkward official title of 'Fighting Falcon'. *(USAF)*

BELOW Equipped with wall-to-wall AIM-9P Sidewinders, this 1981 photograph of an F-16B shows the new horizontal tails (a slightly darker grey), and a belly-mounted AN/ALQ-119 electronic countermeasures (ECM) pod is attached to the fuselage centreline. Observant viewers will also note a spin-control parachute container is attached to the aircraft's empennage, indicating that this is a flight test aircraft. *(USAF)*

ABOVE **Few aircraft scream 'sexy' in the same way as the F-16. With external pylons removed, the aircraft's clean lines and futuristic look (for the time) are apparent. This F-16A belongs to the 34th Tactical Fighter Squadron, 388th Tactical Fighter Wing.** (USAF)

BELOW **The F-16 was eventually cleared for use with 600-gallon fuel tanks, dramatically increasing its ferry and combat ranges in the process. The tanks would be jettisoned prior to entering any combat phase of a mission.** (USAF)

never operate a mix of engine types for logistical reasons.

F-16C/D

The F-16C first flew on 19 June 1984 and was delivered to the USAF one month later. The single-seat F-16C and two-seat F-16D are distinguished by an enlarged base, or 'island', leading up to the vertical fin, from which protrudes a small blade antenna. This island was originally intended for the internal airborne self-protection jammer (ASPJ), but the USAF abandoned this internal jammer in favour of continued use of external ECM (electronic countermeasure) pods.

The F-16C/D introduced a wide-angle General Electric HUD (heads-up display), relocated the cockpit-mounted integrated control panel to below the HUD, and installed the Hughes AN/APG-68(V) radar. It also brought with it a 'glass cockpit' consisting of two multi-function displays (MFDs) that support infrared (IR) video; a MIL-STD-1760 data bus/weapons interface allowing the use of AGM-65D Maverick and AIM-120 AMRAAM (advanced medium-

range air-to-air missile); improved fire control and stores management computers; and a Have Quick II anti-jam UHF radio.

The introduction of these changes was phased. Some were installed at the factory and others as part of the Multi-Stage Improvement Program (MSIP) (avionics and cockpit and airframe changes) and MSIP III (further systems installation), aimed at enhancing the F-16's ability to fly and fight at night.

The F-16C/D Block 30/32 debuted in January 1986 and was the first Viper to feature the 'common engine bay' – the internal modifications necessary to accept either the GE or PW engines of the AFE programme. The Block 30 also introduced a new engine intake configuration for the 'hungrier' GE engine – the so-called 'large mouth' intake allows a greater volume of airflow to reach the engine fan, although very early Block 30s featured the 'small mouth' intake of the PW-powered Block 32.

Besides differences in engine intakes and the common engine bay, the Block 30/32 added full multi-target engagement capability for the AIM-120 AMRAAM from 1987 (Block 30B); it also offered additional computer memory and incorporated a raft of offensive and defence avionics upgrades. The first of 733 Block 30/32 deliveries took place in June 1986. This Block is sometimes referred to as MSIP III.

The Mid-Life Upgrade (MLU) Program was conceived in 1989, initially consisting of a two-year study of the most economical and effective means of updating the F-16A/B, but eventually extending to a six-year development phase that began in May 1991. It comprised a host of updates that spanned the full spectrum of the F-16's avionics, cockpit and weapons systems (see Chapter 2, *Variant summaries and Blocks*). By the end of 1997 it was being rolled out across the four EPAF nations, although the United States dropped its plans to apply the MLU to 223 of its own F-16A/Bs.

In 1988 production moved to the newer F-16C/D Block 40/42, referred to by the Air Force as the F-16CG/DG. Totalling 615 airframes by the late 1990s, the Block 40/42 was intended to supersede the Block 30/32, and improvements comprised a range of modifications aimed at improving the F-16's precision strike capabilities, increasing the

Once the US Air Force had agreed to a high-low mix of F-16s and F-15s, it became imperative to develop the Viper's air-to-ground capabilities. Originally conceived as a good weather point defence fighter, the F-16 would grow to become much more than that. In this series of images, F-16As are seen carrying and releasing a range of ordnance: Mk82 500lb bombs [5], Mk84 2,000lb bombs [6]; GBU-8 2,000lb electro-optically guided glide bomb [1]; CBU-58 cluster bombs mounted on multiple stores ejection racks (MSER) [2]; AGM-65 Maverick fire and forget anti-armour missile [3] and [8]; and the first flight with both the navigation and targeting pods of the LANTIRN (Low Altitude Navigation, and Targeting InfraRed for Night system installed. *(USAF)*

LEFT The first F-16C/Ds assigned to the US Air Force Europe replaced the F-4E Phantom II (left) of the 512th Tactical Fighter Squadron, 86th Tactical Fighter Wing, in September 1985. The Wing was based at Ramstein AB, Germany, using the tail code RS. *(USAF)*

strength of the airframe, and expanding the 9g flight envelope. With an emphasis on night operations, this Block is often referred to as the Night Falcon.

With the end of the Cold War the F-16 ADF operated by select ANG Wings was increasingly becoming surplus to requirements, and in the face of budget cuts it was soon retired from service. The surplus ADF airframes were heavily marketed to foreign operators as an 'off-the-shelf' air defence solution, and several nations purchased these aircraft (see Chapter 3, *International Vipers*).

Production of F-16C/D Block 50/52 followed in 1991, and the type assumed the USAF's suppression of enemy air defences (SEAD) role under the designation F-16CJ/DJ. Importantly, the CJ/DJ featured a new set of Improved

ABOVE By the mid-1980s the F-16 had matured into one of the most capable fighters in the world. The introduction of the C/D model saw the arrival of a wide range of capability upgrades that meant it could perform its multi-role mission even more effectively. This 10th TFS F-16C forms up with an A-10A Thunderbolt and an F-15A Eagle. *(USAF)*

BELOW The F-16C/D also introduced the common engine bay, allowing both the original Pratt & Whitney F100 engine and the new General Electric F110 (pictured here) to be installed. *(USAF)*

BELOW The F-16's 'turkey feathers' can serve as a visual aid to Block identification. These 'feathers' – the 'covers' to the metal louvres that open and close to maintain a constant pressure in the engine exhaust tunnel – are subtly rounded, identifying this as a GE-engined Viper and allowing its identification as a Block 30 (vice Block 32) model. By contrast, the P&W exhaust petals of the Block 32/42/52 series Viper are flat. *(Steve Davies/FJPhotography.com)*

ABOVE The Block 40 F-16CMs (formerly CGs) of the 31st FW, Aviano AB, Italy, give USAFE a vital night-attack multi-role capability. These two 510th FS 'Buzzard' Vipers turn gently above a dense Italian undercast in the skies of Northern Italy. (*Steve Davies/FJPhotography.com*)

ABOVE The F-16CM (formerly CJ) Block 50/52 Viper specialises in the prosecution of enemy threat emitters – a role known as Wild Weasel. Key to the success of this mission is the HARM Targeting Pod, seen here on the right side of this 22nd FS aircraft's intake. (*Steve Davies/FJPhotography.com*)

BELOW Perhaps the most famous operator of the F-16 is the US Air Force's 'Thunderbirds' aerial demonstration team. The team have flown a wide range of Blocks, from 32 to 52. (*Steve Davies/FJPhotography.com*)

BELOW The F-16E/F (F-model pictured) Block 60 Desert Viper is the most capable of them all. Seen here taking off at Nellis AFB, Nevada, during a Red Flag exercise, the aircraft's AESA radar, advanced cockpit and passive IR sensor make it a formidable opponent. (*USAF*)

Performance Engines (IPE) from the two engine manufacturers: F110-GE-129 (Block 50) or F100-PW-229 (Block 52). Both engines are rated in the 29,000lb thrust class.

The Air Force received its final new-build F-16 in March 2005, a Block 50 example that went to the 79th Fighter Squadron, 20th Fighter Wing, Shaw AFB, South Carolina. By 2008 around 700 F-16s remained in active service, 490 equipped the Air National Guard, and 54 were assigned to the Air Force Reserve. Of these, 350 were F-16C/D Block 30s, 51 were F-16C/D Block 32s, 222 F-16C/D Block 40s, 174 F-16C/D Block 42s, 198 F-16C/D Block 50s, and 52 F-16C/D Block 52s.

F-16E/F Block 60 'Desert Falcon'

Representing the most recent member of the F-16 family, the Block 60 is powered by the F110-GE-132 turbofan, offering 32,000lb of thrust in augmentation (afterburner). This variant boasts an integrated electronic warfare suite, an AN/APG-80 active electronically-scanned array (AESA) radar, conformal fuel tanks and a broad range of offensive avionics improvements over the Block 50/52 F-16C/D. Flight testing by Lockheed Martin began in early 2004, with the first Desert Falcons arriving with the type's sole user – the United Arab Emirates, who had ordered 80 – in May 2005.

Chapter Two

Variant summaries and Blocks

F-16s are identified by their model suffix (F-16A, B, C, D, E and F), which identifies the major overall airframe improvements that comprise each variant. They are also identified by their Block number (Block 1, Block 10, Block 25 etc.), which identifies less obvious overall improvements to airframe and avionics fit. In addition, there are also sub-Block suffixes that identify staged improvements within a Block (Block 15Y and 30B, for example), and upgrade programmes that have resulted in improved capabilities and modernisation (the MLU or Mid-Life Upgrade, for example).

OPPOSITE With speedbrakes extended, a Block 25 belonging to the 40th Flight Test Squadron, Eglin AFB, descends over the Gulf of Mexico. *(USAF)*

Because the F-16 has continued to evolve over almost 40 years, the numerous variant and Block numbers listed below provide a graphic illustration of the sheer volume of modifications this has entailed.

It is fair to say that keeping track of all of the differences between the Blocks, upgrades and modifications can be challenging and confusing. Add to this the genuine smorgasbord of upgrade programmes, systems capabilities upgrades (SCUs) and operational flight programme (OFP) software updates, and the picture rapidly becomes confusing. The reality is that the F-16 has now undergone six major Block changes that span four generations of core avionics. Moreover, the introduction of the latest Block brings the number of engine versions to five, while the number of radar

WEASEL VIPERS

It was during the Vietnam War that the USAF introduced the first truly sophisticated suppression of enemy air defence (SEAD) fighters, most notably the F-100F Super Sabre and the F-105G Thunderchief. A rapid proliferation of Soviet-supplied surface-to-air missile (SAM) sites across North Vietnam had rendered the Iron Hand mission, which had typically concentrated on killing anti-aircraft artillery (AAA) emplacements, largely impotent. With specialist mission equipment for hunting and killing SAMs a prerequisite, the Wild Weasel was born. Various iterations of the F-105G were seen, and many lessons were learned, before the archetypal SEAD platform – the F-4G – was finally born. Viewed by many as the most potent SAM killer ever, the F-4G Phantom performed with alacrity during the 1991 Gulf War, following which it was retired promptly from service, much to the chagrin of many a seasoned Wild Weasel pilot. Its successor was the F-16CJ Block 50/52.

The F-16CJ, now known as the F-16CM, is currently employed by the USAF in three main roles, embodying the budget-driven philosophy that each major weapon system (MWS) in the USAF should be able to perform in a truly multi-role capacity: force protection/force projection (SEAD and DEAD), and precision strike role.

SEAD and DEAD are employed against the enemy in either force protection or force projection roles. In the former, Wild Weasel Vipers fly in close coordination with strikers on their way to the target, sniffing out nodes and specific threats within the enemy's IADS and hindering their ability to target the strikers by using the AGM-88 from stand-off ranges, or direct attacks on the offending sites (DEAD) with conventional and smart weapons. The main priority is to get threat emitters off-air as soon as possible; whether simply firing a HARM to force them to cease transmitting, or actually scoring a hard kill against the emitter with a HARM, it matters little. In instances where the threat to the strike force permits, offending sites may be suppressed with HARMs whilst other F-16CMs fly DEAD and close in to permanently neutralise the threat with a hard kill using free-fall munitions. Force projection, on the other hand, concerns itself with proactively striking known nodes of an IADS, or taking out key emitters.

The AGM-88 facilitates both roles with a series of modes of employment. Its most effective mode – and the one that the HTS is geared to provide data for, and makes the F-16CJ unique among other US HARM shooters – is the 'range known' mode, where the missile has accurate azimuth and ranging information. This is the mode that offers the best probability of kill, and known emitter locations can be programmed to the missile prior to flight, or passed dynamically via the ALICS (avionics launcher interface computer) during flight as HTS sniffs the air for electrons. A pre-emptive mode allows the AGM-88 to be fired towards suspected or known sites in a parabolic trajectory that gives it maximum time of flight. In this mode the missile seeker activates as it heads back down towards earth and then waits to see if its assigned target(s) comes on air.

The Wild Weasel Viper (right) replaced the ageing – but much revered – F-4G Phantom II (left). The boots were large, but the Viper would eventually fill them nicely. This scene, photographed at Spangdahlem AB, Germany, shows a mix of the two. HTS was the key tool in making the transition between the two so successful. *(USAF)*

versions and electronic warfare suites has also reached five.

With this in mind, this chapter attempts to outline each variant and Block, describing and summarising as many of their unique capabilities and recognition features as possible. However, it does not claim to provide an exhaustive or complete list of every single feature, addition, retrofit or installation – an entire book could be written on this alone. It also cannot outline the highly modified series of Israeli F-16s, although these are mentioned in brief in Chapter 3.

In addition it should be noted that in recent years the USAF has changed the designations of the C/D-model aircraft. F-16C and F-16D have always been used for Blocks 25, 30 and 32, but the single- and two-seat Block 40/42 night-attack variants were designated F-16CG/DG, while the Block 50/52 Wild Weasel variant was referred to as the F-16CJ/DJ. However, with the 2008 completion of the Common Configuration Implementation Program (CCIP) that ensured Blocks 40/42 and 50/52 shared common avionics equipment, the Air Force consolidated the two designations into the F-16CM (single-seater) and F-16DM (two-seater). Despite this, where it is chronologically appropriate the original designations will be used here for clarity.

Finally, because this volume is primarily concerned with the F-16CM Block 40/42 and Block 50/52 – the most capable and numerous of the F-16s in US Air Force service – these two variants receive the greatest attention.

The text that follows is intended to break down the numerous F-16 models first by variant and then by Block. Update programmes, capabilities and software modifications are discussed at the end of this chapter.

F-16A/B Block 1 to Block 20 overview

The Block 1 to Block 15 F-16 was equipped with a Westinghouse AN/APG-66 pulse-Doppler fire-control radar, a Singer-Kearfott SKN-2400 INS (inertial navigation system), a UHF/VHF communications suite, ILS (instrument landing system), TACAN (tactical air navigation system), a Dalmo Victor AN/ALR-69 RWR

(radar warning receiver), GEC Marconi Avionics HUD and a Sperry central air data computer. All were powered by the F100-PW-200 turbofan, which produced a maximum of 23,830lb thrust. Between 1991 and 1996 the PW-220E engine was introduced, offering additional thrust and an improved lifespan.

The USAF purchased 664 F-16As and 122 F-16Bs up until 1985, when the F-16C/D entered production. Two of the USAF's A-models were built by Fokker in the Netherlands. (See Chapter 3, *International Vipers*.)

F-16A/B Block 1

The F-16A/B Block 1 featured a black radome and RWR covers and was powered by the Pratt & Whitney F100-PW-200 turbofan engine. It had a small UHF antenna below the air intake and was also recognisable by the presence of a smaller vertical stabiliser.

The F-16A was a single-seater, while the F-16B was a two-seat variant. F-16A/B Block 1 aircraft were retrofitted with minor equipment changes and brought up to F-16A/B Block 10 standard under the Pacer Loft programme between 1981 (Pacer Loft I) and 1984 (Pacer Loft II).

A total of 94 Block 1 airframes were produced.

F-16A/B Block 5

The F-16A/B Block 5, of which 154 were built, introduced grey radome and RWR covers in order to make them harder to see in air combat. Powered by the P&W F100-PW-200 turbofan, they were also retrofitted with minor equipment changes. Like the Block 1, they had the smaller vertical stabiliser and UHF antenna below the intake.

ABOVE F-16A 78-064 was a Block 5 Viper delivered in March 1980. This pre-delivery image is noteworthy for a lack of tail codes. It would eventually enter service with the 6516th Test Squadron, Edwards AFB, California. Distinguishing between F-16 Blocks can be very difficult, although the excellent F-16.net website offers an excellent serial-number look-up facility that makes it easy. *(USAF)*

ABOVE From directly above, the slightly smaller horizontal stabilisers of the pre-Block 15 F-16 are apparent. This A-model carries a pair of 600-gallon drop tanks. *(USAF)*

ABOVE RIGHT F-16A 78-0072 was a Block 5 Viper that entered service with 4th TFS in April 1980. Seen here with a single AIM-9L on the right wingtip, this aircraft belonged to 72nd Tactical Fighter Training Squadron (TFTS) at the time it was written off in a landing accident in June 1984. *(USAF)*

BELOW Early USAF Vipers featured coloured warning markings – yellow rescue arrow and red ejection seat warning triangle. These would later be replaced by low-conspicuity markings in dark grey. *(USAF)*

Block 5s were brought to F-16A/B Block 10 standard between 1981 (Pacer Loft I) and 1984 (Pacer Loft II).

F-16A/B Block 10

Block 10 aircraft differed from earlier F-16s through minor internal structural changes. Visually indistinguishable from the Block 5, 355 examples were built. Like the Block 1 and 5 they also had the smaller vertical stabiliser and UHF antenna below the intake.

BELOW LEFT The US Air Force's first operational F-16As are seen here. They entered service with the 388th Tactical Fighting Wing. The aircraft closest to the camera, 79-0359, is a Block 10A that entered service in December 1980. *(USAF)*

BELOW Photographed in soft light, the smooth lines and bubble canopy of F-16B Block 10B 79-430 are evident. The aircraft entered service with the 430th TFS in April 1981, but eventually became a ground instructional airframe – known as a GF-16B – in August 1993. *(USAF)*

Twenty-four Block 10 A- and B-models of the New York Air National Guard (ANG) were modified for the close air support role with the addition of the 339lb General Electric GPU-5/A centreline 30mm gun pod. Developed under the code name Pave Claw, the pod housed a GAU-13/A four-barrelled derivative of the seven-barrelled GAU-8/A cannon installed in the A-10. Whilst Pave Claw was reportedly very accurate, and despite limited use during Operation Desert Storm (1991), integration with the Block 10 was never achieved to the required standard. The 24 NY ANG F-16s were consequently returned to standard Block 10 configuration.

With the introduction of newer F-16 variants over the years, some Block 10 aircraft became GF-16A ground instructional airframes. Others were retrofitted to Block 15 OCU standard (see below) between 1987 and 1993.

F-16A/B Block 15 overview

F-16A/B Block 15 aircraft were built in sub-Blocks as part of the Multi Stage Improvement Program (MSIP). MSIP I was applied on Blocks 15Y to 15AZ, resulting in improvements to air-to-ground and beyond visual range (BVR) air combat capabilities. The first example was rolled out in 1982, although from 1987 all Block 15s were built to Block 15 OCU standard.

MSIP I saw the addition of two small hardpoints (5L and 5R) to the chin of the engine inlet for sensor carriage, as well as the structural strengthening of the chin to support them. This installation shifted the centre of gravity forward, forcing the horizontal stabilators to be enlarged to increase surface area by 30%. These 'big tail' stabilators reduced take-off rotation angle and improved handling characteristics at high angles of attack.

Block 15 F-16s are identifiable by the two parallel RWR antennae beneath the radome and the lack of a blade antenna under the intake. The cockpit layout was also improved.

Block 15 saw important changes to the F-16's avionics suite. A basic track-while-scan capability was added to the APG-66 radar (meaning that the pilot could track one target while continuing to look for others), while the AIM-7 Sparrow air-to-air missile was integrated and approved for carriage and employment. A

ABOVE As the F-16 evolved, a steady effort ensured that new and improved weapons and capabilities were integrated into it. This aircraft, 50-751, carries two GBU-8 glide bombs. Although it is actually a pre-production F-16B, it features the enlarged stabilators of the Block 15 model. Note also the spin parachute housing. *(USAF)*

BELOW Photographed in 1982, this F-16A was operated by an Edwards-based flight test squadron. The shape of the stabilators indicate that this is a Block 15 example. The eight CBU-58 cluster bomb fit was being evaluated for stores separation characteristics. *(USAF)*

LEFT F-16A 81-0741 was a Block 15F example delivered in 1982, and later upgraded to Block 15 ADF standard. It is seen in this 1985 photograph in the markings of the 421st TFS, Hill AFB, Utah. *(USAF)*

LEFT A 119th Wing, North Dakota ANG, F-16A Block 15H ADF launches an AIM-7 Sparrow at a drone at Tyndall AFB, Florida. The Sparrow was cleared exclusively for use by the ADF variant of the Viper, allowing it to reach out and take down bombers that threatened the continental United States. *(USAF)*

F-16A/B Block 15 OCU

Block 15 Operational Capability Upgrade (OCU) aircraft benefitted principally from structural reinforcements and the installation of the enlarged HUD (heads-up display) of the F-16C/D. The OCU improvements were introduced in addition to the Block 15 improvements detailed above, and all Block 15s were manufactured to OCU standard from 1987 onwards, starting with Block 15Y.

The first Block 15 OCU was delivered in January 1988. It introduced an updated radar and associated software, a radar altimeter, a data transfer unit, the AN/APX-101 IFF (identification friend or foe) interrogator and Tracor AN/ALE-40 chaff/flare countermeasures dispensers. The fire control and stores management computers were given additional capacity, and integration with the AN/ALQ-131 jamming pod was completed.

From a munitions perspective, the OCU added the Norwegian Penguin Mk3 anti-shipping missile and the AGM-65 Maverick missile to the F-16's arsenal. Meanwhile, carriage of the AIM-120

ABOVE A nice rear quarter portrait of a Block 15 F-16A. Note the flat 'turkey feathers' of the F100-PW-100 exhaust nozzle. *(USAF)*

Have Quick I UHF secure voice radio system was also introduced.

A total of 984 Block 15 aircraft were manufactured over 14 years, ending in 1996 with delivery of the final example to Thailand. Early USAF Block 15 aircraft were modified to Block 15 OCU specifications between 1987 and 1993.

RIGHT A gaggle of 50th TFW F-16 Block 15s fly formation off the wing of a KC-135 Stratotanker. Meanwhile, the Wing flagship takes on fuel. *(USAF)*

AMRAAM (advanced medium-range air-to-air missile) was also made possible.

With these modifications increasing the F-16's maximum take-off weight to 37,500lb, the 26,660lb (118.32kN) thrust F100-PW-220/220E was also flight tested and approved for use. The PW-220 provides thrust increase while in augmentation of just over 3,000lb.

F-16A/B Block 15 MLU

The MLU is the result of the F-16 remaining in service beyond 1999 – a scenario never originally envisaged. Block 15 Mid-Life Update (MLU) Vipers are refurbished with a cockpit similar to that of the F-16C/D Block 50/52 and are upgraded with the AN/APG-66(V2A) fire-control radar. A miniaturised airborne GPS (global positioning system) has been added, as has a wide-angle HUD, night vision goggles compatibility, a modular mission computer and a digital terrain system.

More details on the MLU programme, which began in 1989 and continues in 2013, are provided at the end of this chapter.

F-16A/B Block 20

F-16A/B Block 20 aircraft are Block 30 airframes built to MLU standards for Taiwan, making these aircraft comparable to Block 50 standard. Historically, F-16 Block numbers had jumped straight from the F-16A/B Block 15 to the F-16C/D Block 25, but Block 20 was created retrospectively. Originally the Block 20 designation applied only to 150 Taiwanese F-16A/Bs; however, it was later used to refer to all MLU aircraft.

Taiwan's Block 20s have the improved AN/ APG-66(V)2 radar of MLU aircraft, but are equipped with an alternative IFF and feature the Raytheon AN/ALQ-183 ECM pod in preference to the Westinghouse AN/ALQ-131.

RIGHT F-16C 83-1129 was a Block 25A Viper assigned initially to the 422nd Test & Evaluation Squadron, Nellis AFB, Nevada. Seen here in three-tone grey, it would later participate in close air support trials, for which it was painted in European One lizard camouflage scheme, consisting of greens and grey. The CAS trials indicated how seriously the expansion of the F-16's mission set was being taken. (USAF)

F-16C/D Block 25 to Block 50 overview

The first F-16C flew in June 1984, commencing flight tests with the US Air Force the following month. The new variant saw the introduction of a radar-reflecting gold pigment in the blown polycarbonate canopy, a feature that helps reduce radar reflections from inside the cockpit.

Other than the tinted canopy, the F-16C differs externally only slightly from the F-16A, its most obvious identification feature being the enlarged triangular base that runs from the rear fuselage up to the vertical fin. A small blade antenna is installed on this so-called 'island'. The enlargement is the result of a planned Westinghouse/ITT AN/ALQ-165 Airborne Self-Protection Jammer, but the Air Force withdrew from the ASPJ programme in January 1990 and the jammer was never installed.

F-16C/D Block 25

The Block 25 F-16C/D entered production in July 1984. In total, 209 C-models and 35

ABOVE In 1986, F-16C/Ds started arriving in Europe, replacing F-16A/B Block 15s. This Hahn-based Block 25F is seen carrying two inert Mk84 dumb bombs over Germany in 1987. (USAF)

D-models were delivered to the USAF, the only user of this Block.

The Block 25 featured the MSIP II improvements that included a Northrop Grumman (previously Westinghouse) AN/APG-68(V) radar. The APG-68 has greater detection ranges, expanded operating modes that include track-while-scan for up to ten targets, improved counter-countermeasures capability and sharper ground-mapping resolution. Amongst other additions, a 'glass cockpit' boasting two multi-function displays, and a wide angle holographic HUD, were also introduced.

The MSIP modifications added even more weight, increasing the maximum take-off load to 42,300lb.

ABOVE A Block 25 F-16C tosses an LJDAM (laser joint direct attack munition) at a target in 2005. This Block introduced an improved radar, wider HUD and new 'glass' cockpit, but relatively few were built. *(Steve Davies/ FJPhotography.com)*

F-16C/D Block 30/32

With Block 30/32 came the reconfigured common engine bay, making it compatible with the new engine to emerge from the Alternative Fighter Engine programme: the GE F110-GE-100, providing 28,984lb (128.9kN) thrust. GE equipped F-16s are designated Block 30, while P&W F100-PW-220 (23,770lb st, 106.05kN) equipped airframes are designated Block 32.

BELOW A nearly brand new F-16C Block 30B of the 13th Tactical Fighter Squadron photographed during exercise Cope North 1987. The Block 30 introduced the common engine bay, but the 'big mouth' inlet for the more 'hungry' GE engine would not feature until F-16C #86-0262 onward. *(USAF)*

BELOW In early 1987, AGM-45 Shrike anti-radar missile testing was done, paving the way for the Block 30/32 Viper to carry the weapon operationally in August that year. Eventually, the AGM-88 HARM – high-speed anti-radiation missile – would also be cleared for carriage. *(USAF)*

Given the GE-100's increased thrust output and correspondingly greater appetite for air, the F-16's inlet was redesigned. Accordingly, early F-16C/D Block 30s have small inlets, but F110-powered Block 30s (F-16C #86-0262 onward) feature 'big mouth' inlets. The official name for this modification is the 'modular common air intake duct'. The Pratt & Whitney Block 32 F-16s all have the smaller inlet, known formally as the 'normal shock inlet'. On the subject of inlets, radar absorbent coatings were applied to both inlets in a bid to reduce the aircraft's radar cross-section.

From August 1987 onwards, Block 30/32 aircraft had the ability to carry AGM-45 Shrike (no longer in the USAF inventory) and AGM-88A HARM (high-speed anti-radiation missiles), as well as the AIM-120 AMRAAM.

Full multi-target capability for the AIM-120 AMRAAM was added in spring 1987, prompting the sub-designation Block 30B. Expanded memory was provided for the programmable display generator and the data entry electronics unit, and the Block 30/32 also introduced the 'Seek Talk' secure voice communication system. Seal-bond fuel tanks were also added. Installation of a voice message unit and crash-survivable flight data recorder rounded off the avionics updates.

The Block 30D doubled the number of ALE-40 chaff/flare dispensers and moved the forward RWR antennae to the leading-edge flap. Dubbed 'beer can' antennae because of their resemblance to aluminium beer

ABOVE Air Force Systems Command's Armament Division conducted extensive testing to integrate the F-16 with the new AIM-120A advanced medium range air-to-air missile (AMRAAM), culminating with full multi-target capability in spring 1987. The resulting Viper was given the sub-Block number 30B. *(USAF)*

containers, these have since been retrofitted to all F-16C/Ds.

Block 30/32 production totalled 706 aircraft, comprising 565 F-16Cs and 141 F-16Ds. The improvements made to the aircraft for this Block are sometimes collectively referred to as MSIP III.

F-16N – Navy Vipers

The Block 30 was the basis for the F-16N and two-seat (T)F-16N – the US Navy's version of the aircraft used exclusively for adversary training. It featured a strengthened wing and, although based on the small-inlet Block 30 F-16C/D airframe, retained the APG-66 radar

LEFT The arrival of the F-16C in Europe, and ongoing efforts to transform it into a Wild Weasel, signalled the end of the road for the F-4G Phantom II. The G-model F-4 was a potent SAM hunter, so the Viper would have big shoes to fill. Here, the old 52nd TFW steed leads the new one. The 52nd TFW was and remains based at Spangdahlem Air Base, Germany. *(USAF)*

US NAVY ADVERSARY VIPERS

The US Navy remains a user of the F-16 in 2013, having first operated the type in 1987. Initially the Navy received 22 F-16Ns and four (T)F-16N two-seaters (based on the F-16C and D respectively), both specifically designed for use in the adversary training role. They featured a strengthened airframe to cope with the constant high-g manoeuvring of dissimilar air combat training (DACT).

The F-16N and (T)F-16N were based on F-16 Block 30E models, and were built to order over the course of 1987–88. VF-126 Bandits, NAS Miramar, achieved IOC in April 1987 aircraft. Then followed VF-45 Blackbirds, NAS Key West; VF-43 Challengers, NAS Oceana; and the Navy Fighter Weapons School, NAS Miramar. In the years that followed, the F-16Ns were subjected to extensive and unrelenting high-g use, resulting in structural fatigue and their early withdrawal from service in 1994.

However, the loss of the F-16 was keenly felt by the instructors teaching F-14 and F/A-18 pilots how to fight dissimilar aircraft, so 14 embargoed Pakistani F-16 Block 15s were taken out of storage and given to the Navy.

Today (June 2013) these same Vipers are still in use as adversaries, but are now flown only by the Naval Strike Air Warfare Center (NSAWC), NAS Fallon, Nevada. NSAWC combines a number of Navy schools that were independent during the 1990s, the most famous of which is Top Gun.

Today, the US Navy's F-16s are based at the Naval Strike and Air Warfare Center, Naval Air Station Fallon, Nevada. Naval Air Systems Command released a fact sheet in October 2012 stating that the inventory consisted at that time of ten F-16As and four F-16Bs. Note the range of camouflage patterns, intended to simulate adversary schemes from around the world. *(USAF)*

LEFT Six Block 42s of the 17th Tactical Fighter Squadron, 20th TFW, Shaw AFB, South Carolina, queue to receive fuel from a KC-135 Stratotanker over Utah in 1989. This version of the F-16 was developed specifically to hunt and kill at night. *(USAF)*

of the F-16A/B. They had their M61-A1 Vulcan cannon removed and are not equipped with missiles, although air combat manoeuvring instrumentation pods (ACMI) were carried. The aircraft were in service between 1988 and 1998.

Production totalled 22 F-16N airframes and four (T)F-16N airframes. Hairline cracks were discovered in the bulkheads of some F-16Ns in the 1990s, but the Navy retired the aircraft in preference to having them repaired. By 2003 they had been replaced by embargoed ex-Pakistani F-16s.

F-16CG/DG Block 40/42 (now designated F-16CM)

Block 40/42 F-16s began to come off the production line in December 1988, with 699 being manufactured in total.

This version was dubbed the Night Falcon in deference to the special night attack capabilities that it featured. At the heart of these were the AAQ-13 and AAQ-14 LANTIRN (Low Altitude Navigation, and Targeting InfraRed for Night) navigation and target pods respectively. A Navstar GPS navigation receiver was also added, so too was compatibility with the AGM-88 HARM II, the improved APG-68V radar, digital flight controls, automatic terrain-following coupled with LANTIRN and, as a

consequence, yet another increase in take-off weight to 42,300lb.

F-16C Block 40 is also powered by the General Electric F110-GE-100, while F-16C Block 42 Falcons are fitted with the Pratt & Whitney F100-PW-220. The Block features 'big mouth' and 'small inlet' versions in the same way as the Block 30/32.

The Night Falcon's airframe was strengthened to increase its 9g capability from 26,000lb (12,201kg) to 28,500lb (12,928kg). The heavier all-up weight, in addition to the installation of LANTIRN, necessitated the installation of a stronger and larger main landing

BELOW The Block 40 F-16C/D began rolling off the production line in late 1988. This F-16D Block 40F was delivered in January 1991, serving with a range of units before being assigned to the 445th Flight Test Squadron, Edwards AFB, California, in January 2008. *(USAF)*

TECHNICAL SPECS: F-16CM BLOCK 50+

Fuselage length: 49ft 4in (15.03m).
Wingspan with tip-mounted AAMs: 32ft 9¾in (10.00m).
Wing area: 300.00ft² (27.87m²).
Wing aspect ratio: 3.0.
Tailplane span: 18ft 3¾in (5.58m).
Vertical tail surfaces: 54.75ft² (5.09m²).
Height: 16ft 8½in (5.09m).
Wheel track: 7ft 9in (2.36m).
Wheelbase: 13ft 1½in (4.00m).

PERFORMANCE

Maximum level speed and altitude: Mach 2.05 at 40,000ft (1,353mph).
Maximum level speed at sea level: 795kt (915mph; 1,472kph).
Maximum rate of climb at sea level: 50,000ft (15,240m) per minute.
Service ceiling: more than 50,000ft (15,240m).
Combat radius: 340 miles (579km) on a high-low-high mission with six 500lb (226kg) bombs.

ARMAMENT

One internal M61 Vulcan 20mm cannon; maximum ordnance of 15,200lb (6,894kg) on one fuselage pylon and six underwing pylons. 5.5g manoeuvring limits: fuselage hardpoint stressed for 2,200lb (998kg), inboard pylons for 4,500lb (2,041kg), centre for 3,500lb (1,587kg) and the outboard for 700lb (318kg). 9g manoeuvring limits: 1,200lb (544kg), 2,500lb (1,134kg), 2,000lb (907kg) and 450lb (204kg), respectively.

An F-16CM Block 50. *(USAF)*

gear, a set of bulged landing gear doors to accommodate larger tyres and wheel hubs, and the consequent relocation of landing lights to the nose gear door.

A pressure breathing system known as Combat Edge was added in order to improve g-tolerance, whilst an enhanced envelope gun sight, a moving ground target bombing capability and many other improvements seen in the Block 50/52 (see below) were also incorporated.

F-16CJ/DJ Block 50/52 (now designated F-16CM/DM)

In December 1991 General Dynamics began delivering the F-16C/D Block 50 and 52. Block 50/52 Falcons introduced the Westinghouse AN/APG-68(V)5 radar with improved avionics computer, but these have since been replaced by the (V)7, (V)8 and (V)9. Other additions to Block 50/52 included a Tracor AN/ALE-47 chaff/flare dispenser, ALR-56M radar warning receiver, Have Quick IIA radio, Have Sync anti-jam VHF radio, full AGM-88 HARM integration and a wide-angle HUD.

A GPS navigation system, ring laser gyro inertial navigation system (RLG INS), horizontal situation display (HSD) – a digital moving map over which flight, weapons, target and navigation data can be laid – and Night Vision Goggle-compatible cockpit lighting were also added. Block 50/52 aircraft ordered after 1997 were delivered with colour cockpit displays and a modular mission computer.

Block 50 F-16s are powered by IPE versions of GE and P&W engines, the 29,588lb (131.6kN) F110-GE-129 (Block 50) and 29,100lb (129.4kN) F100-PW-229 (Block 52) respectively.

All but the oldest USAF F-16C/D Block 50/52 aircraft were later modified to Block 50/52D standard: provision was added for the ASQ-213 pod to be carried under the starboard side of the intake. The ASQ-213, known also as the HARM targeting system (HTS), is the key system that allowed the type to take over the Air Force's suppression of enemy air defences (SEAD) role from the McDonnell Douglas F-4G Phantom in 1992.

HTS uses reprogrammable software loads that permit it to maintain a library of up-to-date

RIGHT The Joint Helmet Mounted Cueing System (JHMCS), seen here attached to the pilot's helmet, provides flight, navigation, threat, radar and weapons aiming cues to pilots regardless of where they are looking. Crucially, it also allows off-bore sight engagements using the AIM-9X Sidewinder. *(USAF)*

threat emitter signatures and to employ new techniques as they are introduced. Revision changes occur as both hardware and software updates are made available.

HTS has its own display page that can be called up on to one of the two Multi-Function Displays in the cockpit, displaying threat systems as symbols in a top-down format. By controlling the field of view of the pod it can be made to scan certain frequencies and sectors of terrain faster or slower, depending upon whether the locations of target emitters are already known. Slower scans will be provided when a large field of view must be monitored; fast scans will naturally occur when

LEFT AND BOTTOM Covered with PACER GEM paint, and featuring PACER MUD radar absorbent materials, these two images of Spangdahlem-based Block 52s clearly show how the paint and RAM coatings fade to a dull matt finish over time. When freshly applied, the paint has a metallic sheen. Even a small reduction in radar and IR signature may make the difference between getting to the merge or not. Whilst a potent threat beyond visual range, the Viper is most dangerous at close quarters. *(USAF)*

RIGHT AF90-809, a Block 52A F-16CM assigned to the 422 Test and Evaluation Squadron, Nellis AFB, Nevada, poses for the camera against a backdrop of mountainous Nevada terrain. Note the HTS pod on the port chin pylon. *(Steve Davies/ FJPhotography.com)*

BELOW Whilst the AGM-88 HARM gives the Wild Weasel both a soft- and hard-kill capability, the AGM-65 Maverick (pictured here being fired from a Block 30C of the 116th FS) allows the Viper to close in on a threat system to ensure its destruction. *(USAF)*

HTS – AN/ASQ-213A HARM TARGETING SYSTEM

HTS is a small pod carried on the right side of the intake cheek that is used to find, classify, range and display threat emitter systems to the pilot. Supplementing the HTS is the AN/ALR-56M advanced radar warning receiver (RWR), AN/ALQ-131(V)14 electronic countermeasures (ECM) pod, and the addition of under-wing-mounted chaff dispensers to complement those already mounted on the lower rear fuselage adjacent to the horizontal stabilisers.

The Manned Destructive Suppression programme funds the development of the USAF's SEAD and DEAD efforts, and adds capabilities to the aircraft and HTS by means

BELOW Pulling its nose above the horizon, this Block 50 F-16CM from the 52nd Wing, Spangdahlem AB, Germany, shows how the HARM can be lofted to extend its range. *(USAF)*

BELOW A massive rocket plume emanates from an AGM-88 HARM following launch from this 'Spang' F-16CM Block 50. The HARM can be launched in both reactive and pre-planned modes. *(USAF)*

of software and hardware upgrades. For example, the HTS R7 (release 7) upgrade was introduced to service in May 2007, but had entered development in 2000 to address evolving threats, allow multi-ship cooperative ranging and provide a precision geolocation capability with which to employ precision weapons in the DEAD role.

An R7-equipped Viper can target joint direct attack munitions (JDAMs) and other GPS-guided weapons against threat systems based solely on coordinates generated by the HTS pod. It can also share these 'GPS quality' coordinates with any other aircraft equipped with the Link-16 Data Link. These 'tight' coordinates mean smaller target location errors and minimise the impact of enemy countermeasures. Similarly, R7 also improved the speed and accuracy of the HTS pod's emitter targeting solutions, allowing improved performance when tasked with the destruction of not only fixed sites, but crucially against mobile threats too. Less obvious benefits include the ability to form a more accurate electronic order of battle picture.

As the R7 programme progressed, testing also focused on the dual-carriage of the HTS and the AN/AAQ-28 Litening II or Sniper ER IR target pods. The USAF initially opted to delay the dual-carriage programme in a bid to prevent the CCIP programme from slipping, but both pods have now been integrated effectively. The addition of the target pod provides the Wild Weasel with a crucial tool with which to supplement the HTS and its other avionics systems. Not only does it facilitate in the primary SEAD/DEAD role – by allowing visual confirmation and laser-guided munition targeting against a threat handed down by the HTS – but it also makes the F-16 a more effective platform in the tertiary roles of precision strike and force projection.

The F-16CJ's ability to react so swiftly to pop-up threats is almost entirely down to clever software programming and the hugely successful integration of HOTAS switchology in the cockpit. Captain Gene 'Owner' Sherer gave an example: 'The jet's smart enough that if we see something in the HTS pod, all we do is put our cursors over the threat and get a pretty accurate idea of its position that will be good enough to launch a missile at it. At that point

ABOVE LEFT The AN/ASQ-213A high-speed anti-radiation missile (HARM) targeting pod (HTS) is used on F-16 Block 50/52 Vipers to engage targets faster and more accurately when tasked with suppression of enemy air defences (SEAD) missions. HTS is able to detect, locate and identify ground-based emitters and assist in targeting GPS-based weapons on those emitters. *(USAF)*

ABOVE RIGHT The interface between the HARM and the HTS is the ALICS – contained within the pylon itself (seen open here). *(USAF)*

we designate the target and let the HARM go. How do we physically do that? We move the cursor switch on the throttle with our left thumb and use our right thumb on the designator management switch and target management switch – the DMS and TMS – which are on the stick. So, we DMS forward to make the hat [HTS cursor] the sensor of interest, TMS forward to designate that target, which will hand it off to the HARM, and then hammer down on the pickle button to unleash the HARM.'

LEFT With the HTS providing accurate GPS-quality coordinates of the threat, the target can be passed off to any of the Viper's offensive sensors. In this photograph, a HARM streaks away with the emitter's location firmly lodged in its computer memory. *(USAF)*

the field of view is very narrow. With a given threat emitter detected, the pod can then be commanded to generate (via the GPS and RLG INS) the threat's geographic coordinates. The 'tightness', or accuracy, of these coordinates depends upon the specific frequency on which the emitter is working.

HTS also ties in with the radar warning receiver display, providing front hemisphere detection across a range of bandwidths. The system is passive, meaning that it operates in a receive-only mode. HTS allows the pilot to select a threat emitter and hand-off the relevant data directly to the AGM-88, meaning that reaction time to pop-up threats is greatly reduced. In doing so, it allows the pilot to cue the AGM-88B/C HARM to specific threat systems.

The CJ uses an avionics launcher interface computer (ALICS), which resides in the AGM-88 HARM launcher pylon, to act as the conduit between the HTS, Central Computer and the missile itself.

The Block 50/52+ added provision for the

joint direct attack munition (JDAM), a GPS-guided, all-weather kit that is installed on 500lb, 1,000lb and 2,000lb 'dumb bombs'. A missile warning system provides a passive capability for the detection of missile launches against the aircraft, while compatibility with 600 US gallon fuel tanks was added, as were conformal fuel tanks (CFTs).

An on-board oxygen generating system (OBOGS) was installed, and a terrain reference navigation system was introduced to improve terrain avoidance and reduce the risk of controlled flight into terrain. Other avionics improvements included the addition of the AN/APX-113 advanced IFF, the joint helmet-mounted cueing system (JHMCS), an ASPIS (advanced self-protection integrated suite) internal countermeasures suite, and the APG-68(V)9 radar. The latter improves detection ranges in the order of 30% whilst increasing processing speed and memory. Importantly, the radar also introduces a synthetic aperture radar mode with a resolution of 2ft.

BELOW Conformal fuel tanks – as seen on this F-16E Block 60 Desert Falcon – are installed between the upper wing and fuselage. They increase fuel capacity whilst freeing up the wing hardpoints for weapons carriage. They are often seen equipping Block 52+ Vipers, but have also been fitted to earlier Block aircraft by foreign operators. *(USAF)*

With the Block 50/52 production line still open as of 2013, a total production number cannot be given. Production on 1 June 2013 stood at 389 Block 50 and 548 Block 52 airframes.

F-16E/F Block 60 Desert Falcon

Developed using $3 billion in funding from the United Arab Emirates, the F-16E (single-seat) and F-16F (two-seat) represent on paper the most capable Vipers to date.

Equipped with CFTs that are similar to those of the Block 50/52+, the Desert Falcon is most readily distinguishable from its forebears by the small optical dome of the Northrop Grumman AN/ASQ-32 IFTS (internal FLIR and targeting system) mounted in front of the canopy. Behind the radome sits the Northrop Grumman AN/APG-80 AESA (advanced electronically-scanned array) radar, which offers more power, increased range and better reliability than the mechanical APG-68 series. The AESA radar can perform simultaneous ground and air scan, tracking, and targeting. Importantly, its agile beam offers a low probability of interception – ie it reduces the chances of detection by threat warning receivers.

To complement the advanced offensive avionics pairing of the IFTS and APG-80, an integrated electronic warfare system (IEWS) provides both advance warning capabilities and automatic countermeasures release. In the cockpit, JHMCS provides visual flight, navigation, offence and defence cues directly on to the pilot's visor, while an advanced mission computer fuses sensor and weapon data that is then displayed on three 5in x 7in colour displays in the cockpit. The F-16E/F's core computer suite has over 2,000 times the memory, and over 260 times the throughput, of the original production F-16.

The F-16E/F flew for the first time in December 2003. By 2008, 55 E-models and 25 F-models had been delivered to the United Arab Emirates, the sole customer for this Block at the time of writing.

ADF variants

The Air Defense Fighter variant of the F-16A Block 15 equipped USAF Air National Guard (ANG) units during the final years of the Cold War. Working closely with F-15 units, the F-16A ADF was responsible for the air defence of the

The MJ tail code [left] on this Block 30B vertical stabiliser identifies it as belonging to the 432nd TFW, then based at Misawa AB, Japan. Note the difference between the small bulge in the 'island' for the F-16C, the much longer and larger bulge of the F-16A Block 15 ADV [below], and the complete lack of bulge for the Block 20 F-16B [bottom]. *(USAF)*

northern United States during that time, but with the subsequent demise of the Soviet Union they were later sold to foreign operators. The ANG retired the last of its F-16As in June 2007.

The ADF conversion was centred primarily on upgrading the existing AN/APG-66 radar to improve small target detection and to provide the continuous-wave illumination that was required to support the AIM-7 Sparrow air-to-air missile.

Externally, the ADF variant's most obvious aid to visual identification was the long, thin, horizontal bulges on the base of its vertical tail. These were the result of moving two flight control accumulators so that the Bendix-King AN/ARC-200 high frequency single-sideband radio could be installed in the leading edge of the fin. Two sets of four-blade 'bird slicer' IFF antennae immediately in front of the canopy and under the chin intake, and a night identification light in the port forward fuselage, also assist in identifying this version.

The ADF featured an advanced IFF; improved ECCM (electronic counter-countermeasures); and provision for GPS and AIM-120 AMRAAM missile data link.

F-16 upgrades

In addition to the development of Blocks and sub-Blocks, there have been numerous upgrades applied to US and international F-16s. These have further enhanced the Viper's capabilities in order to both maintain a qualitative edge over adversaries, and to prolong the type's fatigue life. Systems capabilities upgrades (SCU) and operational flight programme (OFP) software updates also pepper the F-16 upgrade landscape.

OFP (operational flight programme)
OFPs occur on a cyclical basis and have a direct bearing on the capabilities of the aircraft. By way of example, OFP updates in the last decade have included OFP M2+, which allowed the F-16 to carry the MIL STD-1760 BRU-57 smart weapons bomb rack; M3.1+, which provided the aircraft with the software necessary to communicate with an IR target (TGP) and to utilise air-to-air interrogator hardware; M3+, which brought AIM-9X compatibility; and M4+, a much-anticipated

upgrade that allowed the F-16CJ to carry both the HTS pod and the AAQ-28 Litening II and AAQ-33 Sniper XR target pods.

MLU (Mid-Life Upgrade) and PACER SLIP
The MLU is a far-reaching upgrade that significantly improves the Viper's capabilities. However, airframes earmarked for the MLU must first go through the PACER SLIP airframe modification programme. This inspects and repairs hairline cracks that have developed in bulkheads of the aircraft, adding an extra 5,000 hours' fatigue life (flying time) to the airframe in the process.

Central to the MLU is a modular mission computer (MMC). This acts as the 'brains' of the Viper, enabling new capabilities, sensors and weapons to be added. Combined with a new signal data processor for the APG-66(V2) radar, the MLU F-16 can use track-while-scan mode to track up to ten targets whilst simultaneously engaging six with AIM-120 AMRAAM. The radar also has an improved detection and tracking range in the order of 25%. An advanced APX-111(V)I advanced IFF completes the most noteworthy additions of the new avionics package.

Less prominent are the additions of a new airborne GPS receiver, a digital terrain system designed to improve awareness at low altitude, and an improved data modem (IDM) to pass and receive targets over the Link 16 data network. An electronic warfare management system (EWMS), developed by Terma Elektronik AS in Denmark, has been added, as have provisions for both a microwave landing system and domestically manufactured recce pods.

In the cockpit, the MLU Viper features a wide-angle HUD, an advanced display generator, an audio and video recorder, JHMCS, colour MFDs, and updated throttle and stick grips that enhance hands-on throttle and stick (HOTAS) functionality. To this end, an F-16 MLU cockpit closely resembles a Block 40/50 cockpit.

Finally, the software that runs many of the core MLU systems has been continually upgraded. The M3+ and M4+ software installed in MLU jets in 2003 and 2007 respectively adds support for AIM-9X, AGM-154 JSOW (joint stand-off weapon), a helmet-mounted sight, automatic target hand-off to AGM-88 HARM, and various other capabilities. The M5+ and M6+ upgrades arrived subsequently, bringing improved IFF capabilities, a wider range of air-to-ground munitions (GBU-39 small diameter bomb and GBU-54 Laser JDAM), as well as compatibility with the AIM-120D AMRAAM missile.

MLU was applied to 90 Belgian Block 15s; 28 Danish Block 10s and 38 Block 15s; 8 Dutch Block 10s and 132 Block 15s; and 24 Norwegian Block 10s and 33 Block 15s. In

With successive marks/variants and Blocks has come a transformation in the F-16's cockpit layout and man-machine interface. [Top] The F-16A/B (F-16B shown here) featured a radar scope/HUD repeater (right) and an electronic stores panel (left). The F-16C/D Block 25 introduced a new glass cockpit [far left], boasting two multi-function displays (MFDs) on which a range of system, navigation, weapons and attack information can be displayed. The Block 40 added a wide-angle HUD [left] to the mix, with the rear cockpit (shown here) provided with a large HUD repeater. (Steve Davies/FJPhotography.com and USAF)

addition, Portugal, Pakistan and Jordan also put F-16s through the MLU.

Some 449 (situation on 1 June 2013) Block 10 and Block 15 F-16s are expected to go through the programme. The resulting designation is Block 20 MLU.

Sure Strike and Gold Strike

In 1995, 38 F-16C/D Block 40 aircraft of USAFE's 31st Fighter Wing, Aviano AB, Italy, were equipped with Sure Strike. This quick-reaction update consisted of making the cockpit night vision goggles (NVG) compatible and installing an improved data modem (IDM), giving the aircraft quick reaction capability for CAS (close air support) missions being flown over Bosnia.

The IDM allowed the aircraft to receive latitude, longitude and elevation of a target direct from a soldier or controller on the ground. Sure Strike would later be incorporated into OFP M5, and the IDM is now standard on F-16CM and MLU aircraft.

Gold Strike provided an upgrade to Sure Strike in 1997, adding two-way imagery transmission and enabling the pilot to receive and transmit video images in the cockpit.

Falcon STAR, Falcon UP

Tangentially related to SLIP is the USAF-managed Falcon structural augmentation roadmap (Falcon STAR), a structural modification programme that commenced in 2008 and provided structural modification kits to the four European F-16 MLU nations and later the USAF and other international operators.

STAR replaces known life-limited components in a bid to ensure that these airframes will meet their forecast life expectancy of 30 years. The Royal Netherlands Air Force refer to it as PACER AMSTEL (After-MLU Structural Enhancement of Lifetime). Falcon STAR is added to all new-build F-16s at the time of production. Falcon STAR is expected to be completed in 2014.

Falcon UP is a Structural Improvement Program (SIP) that incorporates several major structural modifications into one overall schedule. Affecting all USAF F-16s, it permits Block 25/30/32 aircraft to meet a 6,000-hour service life, and confers an 8,000-hour service life on the Block 40/42. Israel has also put its Viper fleet through SIP.

Starting in April 2007, Turkey's state-owned TUSAS Aerospace Industries put 17 of the Royal Jordanian Air Force's Block 15s though all three structural upgrades, having previously completed Falcon UP modifications on 134 F-16C/Ds of the Turkish Air Force between 1994 and 1999.

RIGHT Despite being designed for night operations, the Block 40 Viper was still put through a number of quick capability upgrades during the Balkans conflicts of the 1990s in order to improve its night strike and close air support capabilities. *(USAF)*

HAVE GLASS

HAVE GLASS is the collective name given to a two-stage programme that reduces the F-16's IR and radar cross-section by around 15%. HAVE GLASS I saw the introduction of the gold-tinted metallised canopy that reduced radar reflections from the cockpit. In addition a foam panel was added behind the radar antenna.

HAVE GLASS II sees the application of PACER MUD radar absorbent materials (RAM) to the inlet duct and other elements of the airframe. The RAM consists of ferromagnetic particles embedded in a high-dielectric-constant polymer base. PACER MUD covers about 60% of the airframe (forward and side panels) to a depth of 10-12mm, on top of which a topcoat of PACER GEM paint is applied to reduce the aircraft's IR signature.

HAVE GLASS was first applied to USAF Block 50 Vipers in 2005 and has since been applied more widely, including on the F-16s of the four European Air Forces. Around 1,700 F-16s are expected to receive the HAVE GLASS treatments.

CCIP and CUPID

The Common Configuration Implementation Program has given the USAF's F-16C/D Block 40/42 and Block 50/52 fleet core commonality with both each other and the European MLU. CCIP, which began in 2002 (although some sources state 1998), was split into three main phases: Phase I and IA for introduction of the initial systems to Block 50/52; Phase II for full modification to Block 50/52; and Phase III for full modification to Block 40/42.

Phase I installed the MMC, the APX-113 advanced IFF and the BRU-57 smart bomb rack (increasing the number of GPS weapons the F-16 can carry). It also gave the Block 50/52 the capability to carry a chin-mounted AN/AAQ-33 Sniper XR target pod in addition to the HTS, whilst installing colour MFDs in those Vipers not equipped with them. Phase II added the advanced Link 16 data link, JHMCS, and an electronic horizontal situation indicator. From 2001 to 2006, the Ogden Air Logistics Center at Hill Air Force Base modified and delivered 254 USAF Block 50/52 aircraft, 100 of which had been modified twice in order to complete Phase I and Phase II, and the remainder of which had been modified in a single visit.

Phase III, which began in 2005, applied the entire update to individual Block 40/42 airframes in a single cycle. It also saw Block 40/42 aircraft undergo Falcon STAR at the same time. In 2010, the 500th upgrade was completed.

CCIP modified aircraft are designated F-16CM and F-16DM, and once loaded with OFP tape M4.2 can use the same Modular Mission Computer Software tapes.

The ANG ran a comparable, yet smaller, upgrade programme on its fleet of Block 30 Vipers. Completed by the end of 2003, CUPID – Combat Upgrade Implementation Details – added LANTIRN and Litening II capability, installed a GPS/INS, NVG lighting, a situational awareness data link (SADL), the ALQ-213 countermeasures dispenser, and introduced compatibility with GPS guided weapons. 620 airframes were updated, and the ANG sometimes refers to these as the F-16C+.

SLEP and CAPES

By 2011 the USAF was considering a Service Life Extension Program (SLEP) to the newest Block 40/50 F-16s in order to reach a 12,000-hour fatigue life. Structural upgrades will extend airframe structural service life by 25% (6–8 years), but also repairs the bulkhead cracks in approximately 67% of Block 40/42/50/52 F-16 aircraft. Figures released for 2011 revealed that 285 F-16s had seen bulkhead cracks repaired, while 83 had had new bulkheads installed. Fifty-four F-16s were subject to more regular inspections to measure crack growth.

By 2012 the F-16 SLEP was slated to include up to 350 Block 40/42/50/52 airframes, with the first kits scheduled for procurement in fiscal year 2017 and installations beginning in FY2018.

SLEP has also grown in scope to include the CAPES upgrade. The significant delays in entry to service of the Lockheed Martin F-35 Lightning II has forced the Air Force to consider putting the SLEP airframes through a new combat avionics programmed extension suite (CAPES) upgrade in order to bridge the capability gap. CAPES includes an AESA radar, a Terma AN/ALQ-213 electronic warfare system, an integrated broadcast system (IBS) and a centre display unit (CDU).

Chapter Three

International Vipers

This chapter briefly outlines the international operators of the F-16. Listed alphabetically, starting overleaf, they are (see overleaf):

OPPOSITE The F-16E is the latest version of the Viper. The E/F designations were originally reserved for the production version of the F-16XL – a cranked, delta-wing Viper design that competed with the F-15 in 1980s for the Advanced Tactical Fighter programme. *(USAF)*

Bahrain

The Bahrain Amiri Air Force ordered 12 F-16C/D Block 40s under the Peace Crown I Foreign Military Sales (FMS) programme in March 1987. The first aircraft arrived in Bahrain in May 1990. Peace Crown II saw the delivery of ten more Block 40s in 2000. They are equipped with the AN/AAQ-14 Sharpshooter target pod (the export version of the LANTIRN target pod).

Belgium

As one of the first four international F-16 customers, the initial 1975 Belgian order was for 116 F-16A/Bs, the first Belgian-built F-16 of which was delivered in December 1978. In 1981 35 Block 1/5 airframes were upgraded to Block 10 standard. Then, in 1988, the first of a batch of 44 Block 15 OCU examples was delivered.

Belgium ordered 160 F-16s, all of which were built by Belgian-owned SABCA (Société Anonyme Belge de Constructions Aéronautiques) in two production batches. The last example was delivered in April 1985, making it the final NATO F-16 (totalling 348 airframes). A second order for 44 examples was completed by SABCA in 1991.

Ninety Belgian Vipers have been through the F-16 MLU programme. Belgian Air Force F-16s are equipped with the French-built Carapace ECM package, making them unique amongst F-16 operators.

Belgian defence reorganisation plans in 2003 earmarked the fleet for reduction to only 60 aircraft by 2015. Combined with attrition, this reduction meant that by 2005 its operational inventory had already been reduced to 72 aircraft.

Chile

The Chilean Air Force ordered ten new F-16 Block 50s (six C-models and four D-models) in December 2000 under the Peace Puma FMS programme. Deliveries started in 2005 and were complete by the end of 2006.

Prior to the first Block 50 arriving, the Fuerza Aerea de Chile stated that it would supplement its new Viper fleet with the acquisition of 18 second-hand F-16A/B Block 20 MLU aircraft obtained from the Netherlands (deliveries of which ran between 2006 and 2007 under Peace Amstel I). In 2009, Peace Amstel II, the Netherlands sold Chile an additional 18 F-16A Block 20 MLU aircraft.

Chilean F-16s are devoid of a TACAN navigation system (there are no TACAN transmitters in South America). Whilst some reports state that they are also unable to employ either AGM-88 HARM or AIM-120 AMRAAM missiles, US Defense Security Cooperation Agency reports prepared for Congress indicate that both weapons were sold to Chile in 2009.

Denmark

Denmark has 77 F-16A/Bs, of which 66 are upgraded to MLU standard. Denmark's first 58 Vipers came from SABCA in Belgium, beginning with the delivery of an F-16B on 18 January 1980. These F-16A/B Block 1 aircraft were later upgraded to F-16A/B Block 10 standard by the Aalborg RDAF workshop in the Pacer Loft I programme.

In August 1984 Denmark ordered 12 follow-up F-16s (eight F-16As and four F-16Bs), built by Fokker in the Netherlands. Two attrition replacement batches have since been purchased (three airframes in 1994, four more in 1997).

Danish F-16s feature an identification lamp on the port side of the nose and have Pylon Integrated Dispensers on the outboard wing stations.

Egypt

The F-16 is the primary fighter in the Egyptian Air Force (*Al Quwwat al Jawwiya Ilmisriya*) inventory, numbering 240 Vipers in total.

The first of 42 F-16A/B Block 15s arrived in March 1982 under Peace Vector I (these aircraft were modified to Block 42 standard from 1997 onwards). Peace Vector II followed, seeing the delivery of 40 F-16C/D Block 32s between 1986 and 1988. Next, Peace Vector III resulted in the October 1991 delivery of the 35 F-16C Block 40s and 12 F-16D Block 40s. A fourth order, Peace Vector IV, resulted in TUSAS Aerospace Industries (TAI) of Turkey being contracted to provide 46 licence-built Block 40s, although deliveries had to be made via the United States for legal reasons. These aircraft were delivered between 1994 and 1995.

LEFT Of Denmark's 77 F-16A/Bs, 66 were upgraded to MLU standard. Like other Scandinavian nations, Denmark has no air-refuelling capability of its own, and is therefore reliant on other nations in order to train and remain proficient in this skill. In this image the 100 ARW stencil on the boom indicates that this is a USAFE Stratotanker based out of RAF Mildenhall, England. *(USAF)*

CENTRE AND LEFT The fly-by-wire boom of a KC-10 Extender is held slightly right as an Egyptian Block 15 F-16B approaches in 1983. By 1991 Egypt had also received Block 30 (seen here) and Block 40 Vipers. Note the beer-can RWR antennae on the leading-edge flaps. *(USAF)*

47

INTERNATIONAL VIPERS

ABOVE AND ABOVE RIGHT Greece is the most numerous operator of the Viper in Europe, owning a mixed fleet of Blocks 30, 40, 52 and 52+. Here 015, a Block 52 F-16C delivered under Peace Xenia IV, sits waiting for the next sortie. Meanwhile 024, an F-16D Block 52+ with enlarged dorsal spine and CFTs, cruises over the sea.
(Hellenic Air Force)

Peace Vector V and VI produced 21 F-16C Block 40s from 1999 to 2000, and 24 F-16C/D Block 40s between 2001 and 2002, respectively. The latest order, Peace Vector VII, came in the form of 20 Block 52 aircraft delivered between 2012 and 2013.

Greece

The Hellenic Air Force has received 170 F-16s, including F-16C/D Block 30, 50 and 52, making it Europe's most numerous Viper operator.

Greece ordered 40 F-16C/D Block 30s in January 1987 under Peace Xenia I, with deliveries starting in November 1988. Peace Xenia II, for 40 Block 50s with GE engines, was placed in April 1993, and deliveries began in July 1997. Peace Xenia IV followed in June 2000, consisting of the purchase of 50 brand new F-16 Block 52+ aircraft, to which ten more were added in September 2001. These aircraft are all equipped with conformal fuel tanks, while a dorsal spine is also fitted on the D-models to allow for the installation of additional avionics. Deliveries were complete by June 2004.

The Peace Xenia I Block 30s have received the Falcon UP upgrade, installing the ASPIS electronic warfare system that consists of ALQ-187 jammer, ALR-66VH(I) RWR and ALE-47 countermeasures dispensers. HAF Block 50/52s are equipped with the AN/AAQ-14 target pod. Greece's Vipers carry the IRIS-T air-to-air missile in addition to the AIM-120 AMRAAM.

Indonesia

In August 1986, the Indonesian Air Force (*Tentara Nasional Indonesia-Angkatan Udara*) ordered 12 F-16A/B Block 15 OCU aircraft under the Peace Bima-Sena programme. The first aircraft was delivered in December 1989 and the final delivery took place in 1990.

The TNI-AU signed up to acquire nine additional F-16A Block 15s in March 1996, but in July 1997 the order was cancelled following political tensions between the Indonesian government the the United States. A US military embargo on Indonesia followed, leading to the rapid ageing of the TNI-AU's Viper fleet. The embargo was lifted in 2005, and reports indicate that ten of the 12 original Vipers then remained in operation.

With relations back on track, Indonesia placed an order for 24 ex-USAF Block 25 F-16C/Ds. US confirmation of the order followed in November 2011. These aircraft will be upgraded to Block 52 standard, and deliveries are expected to be completed by July 2014, although by July 2013 none had yet been delivered.

The TNI-AU operate the Viper in both air defence and ground attack roles, but limit operations to daylight hours only.

Iraq

The Iraqi Air Force (*Al Quwwat al Jawwiya al Iraqiya*) placed an initial order for 18 F-16C/Ds in 2011, adding to this in October of 2012 with a follow-on order of another 18 airframes. Deliveries are slated to begin by 2014.

The aircraft will employ the Goodrich Corporation's DB-110 airborne reconnaissance system – a real-time tactical reconnaissance camera enclosed in a 19ft (6m) belly-mounted pod.

Israel

With both the largest and the most eclectic mix of F-16s outside the United States,

Israel purchased no fewer than 362 F-16s between 1980 and 2000. These Vipers were all purchased under successive iterations of the Peace Marble Foreign Military Sales programme. They are operated by the Israeli Defence Force/Air Force (*Cheil Ha'avir*).

The first four F-16s, christened 'Netz' (Hawk)

ABOVE LEFT AND RIGHT Israel has the largest mix of F-16s outside of the United States, having received its first four Vipers in July 1980. Since then the country has equipped its fleet with a range of indigenous weapons and avionics systems. This customisation is epitomised by the F-16I Sufa (shown here), a derivative of the F-16D Block 52. Note the conformal fuel tanks and the use of the locally-produced DASH helmet cueing system in place of JHMCS. *(USAF)*

ISRAELI F-16 OPERATIONS

Entering combat shortly after delivery in 1980, the F-16s of the Israeli Air Force have forged an exceptional operational record in the Middle East, although they are by nature extremely secretive and rarely publicise the details of their exploits.

Israel was a customer of the French Dassault Mirage series through the 1970s, but the peace agreements of the 1980s – particularly the signing of the Camp David Peace Treaty with Egypt – enabled the Israeli Defence Force to purchase arms from the United States.

Before the delivery of the Peace Marble I aircraft was complete, the IAF's brand new F-16A experienced its baptism of fire when, on 28 April 1981, a 'Netz' downed a Syrian Mi-8 helicopter near the Lebanese town of Zahle. Later that year, Operation Opera saw eight F-16As armed with 2,000lb general-purpose bombs dispatch Iraq's Osirak nuclear reactor in what was a pre-emptive, deep penetration strike. The raid covered more than 1,000 miles and was a daring mission that pushed the F-16 beyond what many thought it was capable of. Of the eight F-16s in the strike package, all but one scored direct hits with their Mk84 bombs.

By 1982 the 'Netz' was in the thick of the action once more, this time over the Bekaa Valley, where it notched up an impressive tally of 44 aerial victories during air combat with Syrian MiGs.

Having proved their worth several times over, Israel

requested and received more F-16s in October 1987. These comprised the improved C and D models equipped with a sophisticated Westinghouse AN/APG-68 pulse-Doppler multi-mode radar and General Electric's more powerful F110-GE-100 afterburning turbofan. The new type was called the 'Barak', and in typical IAF fashion it saw its first combat activity soon after deliveries commenced. In April 1988 the new 'Baraks' were used to attack terrorist bases south of Beirut. The targets included command centres, ammunition depots and weapon systems, and the IAF F-16C/Ds accounted very well for themselves.

In more recent times, the highly modified and highly advanced F-16I 'Sufa' was used to conduct a deep penetration strike in 2008 against a Syrian target that, despite being shrouded in secrecy, is believed to have been a nuclear research facility. The IAF Viper fleet also played a significant role in the 2006 and 2008 conflicts in South Lebanon and Gaza respectively, even downing three Iranian-supplied unmanned aerial vehicles loitering near Gaza during the course of 2006.

Israel's F-16s have been at the forefront of its military operations almost since the day the earliest examples were delivered. It is safe to assume that the Viper units of the IDF/AF will continue to play a major role in shaping the future of Israel's political position within the Middle East.

by the IAF, arrived in Israel in July 1980 following the 1978 Peace Marble I. Whilst the Reagan administration blocked deliveries of the last 22 'Netz' following Israel's attack on Iraq's Osirak nuclear reactor, final deliveries did eventually take place in 1981. Eighteen of the Peace Marble I airframes were Block 5s upgraded to Block 10 standard. Peace Marble II followed, supplying Israel with F-16C/D Block 30s. The first of 75 'Baraks' arrived in October 1987.

Following the cancellation of Israel's Lavi fighter project in May 1988, Peace Marble III enabled the acquisition of 60 F-16C/D Block 40 'Barak IIs', delivered from August 1991. With the Gulf War (Operation Desert Storm) petering out, Israel was rewarded for its restraint (in not retaliating against Iraqi Scud attacks) with the hastily-approved Peace Marble IV, consisting of 50 ex-USAF Block 10 F-16A/B 'Netz IIs'.

In September 2001 Israel finalised its purchase of 102 F-16D Block 52s – known as the F-16I 'Sufa' – under Peace Marble V. The purchase, valued at $4.5 billion, saw the first two 'Sufas' delivered to Mizpe Ramon Israeli Air Force Base in February 2004. The remainder were delivered during 2009.

Briefly, the F-16I features the Pratt & Whitney F100-PW-229 engine and has an indigenous avionics suite that ties in with a range of systems: an internally mounted FLIR sensor, Python 5 imaging infrared-guided air-to-air missile, AN/APG-68(V)9 radar and an Israeli electronic warfare suite. Externally, a dorsal compartment houses the avionics hardware in addition to chaff and flare dispensers, while CFTs free-up underwing hard points. The rear cockpit is occupied by a weapons systems officer.

Italy

Under Peace Caesar, the Italian Air Force leased 34 Block 5/10/15 ADF F-16A/Bs as a stopgap measure until the Eurofighter Typhoon entered service. The aircraft were ex-USAF examples taken from storage, all of which had been extensively refurbished. Delivered between 2003 and 2004, they were phased out of service in May 2012.

Jordan

The Royal Jordanian Air Force (*Al Quwwat al Jawwiya al Malakiya al Urduniya*) acquired the first of 18 leased Peace Falcon I F-16A/B Block 15 ADF fighters in 1997. Peace Falcon II followed in 2003, adding 17 additional

BELOW AND RIGHT The Royal Jordanian Air Force originally leased F-16A/B Block 15 ADFs, but would later purchase Block 20 MLUs from Belgium and the Netherlands. In the first image an early ADF Viper takes on fuel. In the second, one of Jordan's Block 20 MLUs leads a mixed formation of (from bottom to top) Pakistani Air Force Mirage, US Air Force F-16 Block 52 and US Navy F/A-18C Hornet. *(USAF)*

examples to the inventory. In 2004 an agreement was reached with Lockheed Martin to put these airframes through the Falcon UP and Falcon STAR upgrades (actually performed by Turkish Aerospace Industries at Ankara), in addition to cockpit updates.

Between 2006 and 2007 eight ex-Dutch Air Force and 14 ex-Belgian Air Force F-16AM/BM Block 20 MLUs were acquired under Peace Falcon III, with deliveries running through 2009. That same year, the ex-Dutch F-16A Block 20 MLU order was cancelled, meaning that Peace Falcon III actually finished with the delivery of only 16 airframes. However, this was offset by the delivery of six ex-Dutch F-16B MLUs that had been ordered under Peace Falcon IV in 2005. Also in 2009, Jordan began negotiations for six F-16AMs and three F-16BMs from Belgium, resulting in the Peace Falcon V programme. Delivery was completed in July 2011, bringing the total Viper acquisition for Jordan to 64.

Morocco

Twenty-four F-16C/D Block 52s were ordered by the Royal Moroccan Air Force (*Al Quwwat al Jawwiya al Malakiya*) in 2008. Deliveries ran between 2010 and 2011.

Norway

Norway, one of the original members of the European Participating Air Forces group, ordered 72 Block 1/5/10/15 standard F-16s from the Netherlands Fokker production line between January 1980 and June 1984, while two F-16B Block 15 OCUs were acquired direct from General Dynamics in 1989 as attrition replacements.

Royal Norwegian Air Force (*Luftforsvaret*) F-16s are equipped with tail-mounted braking parachutes, an identification spotlight, the Northrop Grumman AN/ALQ-162 internally mounted deception jammer upgraded to Shadowbox II standards, and can employ the Penguin anti-ship missile. Fifty-six airframes have been upgraded to MLU standard.

The type is expected to be phased out from 2015, the anticipated delivery date of the Lockheed F-35 Joint Strike Fighter, although delays in that programme may see the F-16 remain in Norwegian service for longer.

Oman

FMS programme Peace A'sama A'safiya I saw the Royal Air Force of Oman (*Al Quwwat al Jawwiya al Sultanat Oman*) acquire 12 F-16C/D Block 50s between 2005 and 2006. Peace A'sama A'safiya II was agreed in August 2010, adding 12 more Block 50s. Deliveries commenced in 2013. Oman also purchased the Sniper target pod, and, like Iraq and Poland, purchased the DB-110 airborne reconnaissance pod.

Pakistan

Pakistan's Viper acquisitions have taken place under the Peace Gate and Peace Drive foreign military sale programmes. Peace Gate I was agreed in 1981, seeing the country order 40 examples of the F-16A/B Block 15, the first six being delivered to the Pakistan Air Force (*Pakistan Fiza'ya*) in January 1983. The remaining 34 Peace Gate I aircraft were actually delivered under the auspices of Peace Gate II between 1983 and 1987.

Eleven F-16A/B Block 15 OCU aircraft were requested under Peace Gate III in December 1988, while in December 1989 Pakistan signed an agreement for the purchase of 60 more Block 15 OCUs under Peace Gate IV. However, while Pakistan had already paid for the 11 Peace Gate III aircraft, the sale of all 71 Peace Gate III and IV Block 15 OCUs was embargoed by the United States in October 1990 when it became clear that Pakistan was pursuing a nuclear weapons programme.

Pakistan's assistance to the US during the so-called global war on terror from 2001 onwards has since resulted in the lifting of that

ABOVE AND BELOW Pakistan has a chequered history with the United States, leading at one point to Vipers that it had already paid for being embargoed and given instead to the US Navy. Today, with a relationship that appears to exist solely in the interests of international stability, Pakistan is receiving the latest F-16s. Note the dorsal spine of the air-to-air photograph of the F-16D Block 52. *(USAF)*

embargo. Consequently a September 2006 approval for the acquisition of 18 F-16C/D Block 52s (with an option for 18 more) was given by the United States. The agreement – Peace Drive – made provision for the delivery of 26 Peace Gate III and IV aircraft, and also provided for the MLU and Falcon UP upgrades for Pakistan's existing F-16A/B fleet. Deliveries began and concluded in 2010.

Pakistani Vipers carry the ALTIS laser target pod and the Magic II air-to-air missile, both of which are French. The Peace Drive Block 52s feature conformal fuel tanks.

Poland

The Polish Air Force began receiving the first of 48 Vipers in late 2006 under Peace Sky. Poland's F-16C/D Block 52+ aircraft boast similar capabilities, with Sniper XR target pods and the DB-110 airborne reconnaissance system complementing the AN/APG-69(V)9 radar. For self-protection, Polish Vipers have the AN/ALQ-211(V)4 system, providing unrivalled protection in high-threat environments.

Portugal

The Peace Atlantis I programme paved the way for the arrival of the first of 20 F-16 Block 15 OCUs to Portugal in July 1994. Peace Atlantis II of 1996 resulted in the acquisition of an additional 25 ex-USAF F-16A/B Block 15s, with deliveries commencing and completed in 1999. Twenty of the Peace Atlantis II airframes have since been refurbished with the Falcon UP structural upgrade, F100-PW-220E engines and the F-16A/B MLU avionics and cockpit upgrade. Another 18 airframes were later added to the MLU programme.

Republic of China (Taiwan)

The Republic of China Air Force (*Chung-kuo Kung Chun*) operates around 150 F-16A/B Block 20 aircraft, forming a key part of the country's defence against the perpetual threat that is mainland China.

LEFT Photographed during the Falcon UP structural upgrade, this Portuguese F-16 is undergoing reassembly prior to a functional check flight. *(USAF)*

Peace Fenghuang was agreed in November 1992, allotting Block 20 F-16s for delivery commencing in 1997. However, in both 2006 and 2009 attempts to purchase F-16C/D Block 52s were blocked by the United States, as were plans to upgrade the Block 20s. In 2011 a letter of request from Taiwan was also blocked, leading to the conclusion that the RoCAF's Viper acquisition has truly stalled.

In 2012 an upgrade package for Taiwan's Block 20 Vipers was at last concluded, and this included retrofitting the AN/APG-66(V)3 radar with an AESA radar and installing software updates for the engine control units.

The RoCAF's Block 20s boast an improved Westinghouse AN/APG-66(V)3 fire-control radar and the ALR-56M advanced radar warning receiver. Equipped with the Sharpshooter and Pathfinder target and navigation pods, the AN/VDS-5 recce pod is also employed.

Singapore

The Republic of Singapore Air Force (RoSAF) is equipped with 62 F-16C/D Block 52 aircraft, some of which are outfitted with the Israeli DASH-3 Helmet Mounted Sight and Israeli electronic warfare equipment.

The country's utilisation of the Viper started in 1988 with the purchase of eight F-16A/B Block 15 OCU aircraft under Peace Carvin I. All eight had been built using Block 30 airframes. Peace Carvin II cleared the way for 18 Block 52s to be acquired from 1998 onwards, ten of which feature the enlarged dorsal spine of the F-16I 'Sufa'. Later 12 additional examples were initially leased from, and then finally purchased outright directly from, Lockheed Martin (these are based at Luke AFB, Arizona, where they are used for RoSAF pilot training).

Peace Carvin III resulted in the delivery of 12 more Block 52 Vipers between 2000 and 2002, while Peace Carvin IV led to deliveries of another 20 Block 52s (with the dorsal spine) by the end of 2004. Singapore transferred its remaining Peace Carvin I F-16A/Bs to Thailand in 2005.

South Korea

The Republic of Korea Air Force (*Han-Guk Kong Goon*) has purchased 180 F-16C Block 32 and Block 52 F-16C/Ds. Peace Bridge I brought about an order for 40 Block 32s, the first of

which was delivered in 1986. Of 120 Block 52 examples ordered under Peace Bridge II, all but 12 were either assembled or manufactured in South Korea by Samsung Aerospace under the designation KF-16. The first KF-16 arrived with the RoKAF in June 1997.

Twenty additional Block 52s were requested under Peace Bridge III FMS, with deliveries from Samsung Aerospace lasting from 2003 to 2004. South Korea is the first international customer to equip its Vipers with the AN/ALQ-165 Airborne Self-Protection Jammer, doing so as part of a modernisation programme that commenced in April 1997.

Thailand

The Royal Thai Air Force (*KongTup Arkard Thai*) has ordered 54 F-16A/B Block 15 OCU and Block 15 ADFs over the course of four Peace

ABOVE A pair of KF-16D Block 52Ds of the South Korean Air Force fly formation with two F-16C Block 52s of the 8th FW. Interestingly, both KF-16s carry live AIM-9M and AIM-120 missiles. Heavy deposits around the M61A1 ports of both aircraft also indicate that gunnery practice has been high on the agenda for South Korean pilots. *(USAF)*

BELOW A Royal Thai Air Force F-16 photographed in 2013. The large protuberance on the vertical tail's 'island' indicates that this is one of Thailand's Block 15 ADF examples. Thailand uses a slightly different serial number convention to most F-16 operators: this aircraft is serialled 10214 (Lockheed Martin serial 81-0784). *(USAF)*

Naresuan FMS programmes, deliveries of which ran between 1988 and 2005.

The first 12 Peace Naresuan I Block 15 OCUs arrived in 1998, while six more followed between 1990 and 1991 under Peace Naresuan II. Peace Naresuan III resulted in 18 additional examples arriving between 1995 and 1996. Later, Peace Naresuan IV created a mixed purchase of 16 Block 15 ADFs and two Block 10 OCUs (the Block 10s being earmarked for cannibalisation for spares) that were delivered in two batches between 2002 and 2003.

In exchange for the use of Thai training bases and airspace over the years to come, Singapore gave its seven Peace Carvin I Block 15 OCUs to Thailand in 2005. Thai Vipers carry the ALTIS II target pod and the Rubis navigation pod.

The Netherlands

In March 1980 the Royal Netherlands Air Force (*Koninklijke Luchtmacht*) placed an order for up to 213 F-16A/Bs Blocks 1–15. The KLu

upgraded 138 of its F-16s to MLU standard, creating the F-16AM/BM. In 2003 the Dutch government cut the F-16 force by 25%, and in 2005 the KLu began selling small batches of its Vipers (first to Jordan, then to Chile).

The Netherlands operates modified recce F-16s that it designates F-16A(R) and F-16B(R); these initially carried the Oldelft Orpheus pod, but eventually adopted the Medium Altitude Reconnaissance System. In 2006, the KLu placed an order for eight Elbit Reccelite pods and Litening AT pods. These were delivered in 2008.

Turkey

The Turkish Air Force has 270 Block 30/40/50 F-16C/Ds in its inventory. Peace Onyx I was the first order, placed in September 1983, for 160 Block 30/40 F-16s. All but the first eight aircraft were assembled in Turkey by Tusas Aerospace Industries (TAI). Deliveries started in October 1987. The Block 40 aircraft are powered by General Electric F110-GE-100 engines built under licence by TAI Engines at Eskisehir.

Peace Onyx I aircraft were fitted with the ALQ-178(V)3 Rapport III ECM system, but this was later upgraded to the ALQ-178(V)5 Rapport III. In 1994 TAI began the Falcon UP modification programme on the Peace Onyx I aircraft, consisting mainly of structural improvements.

In March 1992 Peace Onyx II saw 40 F-16 Block 50s ordered, with deliveries continuing into 1997, and a later upgrade to the ALQ-178(V)5 Rapport III. Between 1998 and 1999 Peace Onyx III resulted in 40 additional Block 50s being delivered, this time with the ALQ-178(V)5 Rapport III installed prior to delivery.

ABOVE AND ABOVE RIGHT Turkey's Vipers regularly duel with those of Greece over the Aegean Sea. Whilst such encounters are usually without incident, there have been losses as a result of aircraft flying into the sea while fixated on air-combat manoeuvring. Here, Turkish Block 30Bs are seen landing and preparing to taxi. *(USAF)*

Thirty more Block 50s were purchased under Peace Onyx IV in 2007, with deliveries running between 2011 and 2012.

In April 2005 the Turkish government confirmed plans to upgrade 217 F-16s (38 Block 30, 104 Block 40 and 76 Block 50) to CCIP standard.

The United Arab Emirates

The United Arab Emirates (*Al Imarat al Arabiyah al Muttahidah*) boasts the most advanced F-16. It has 80 F-16E/F Block 60 Vipers on strength since deliveries were completed after less than two years in 2006.

Venezuela

The Venezuelan Air Force (*Fuerza Aérea Venezolana*) received its 24 F-16A/B Block 15s via Peace Delta between 1982 and 1984. A long-standing arms embargo that has since been in place between the US and Venezuela means that the Peace Delta Vipers are difficult to maintain.

In 2012 Iran stated that Venezuela had given it at least one of its F-16s for exploitation. Rumours abound about Israeli and Belgian assistance to the South American country in maintaining and arming its F-16s, but none has been substantiated.

LEFT AND BELOW LEFT The F-16E/F Desert Falcon equips the air force of the United Arab Emirates. Initial training was conducted at Luke AFB, Arizona, with those UAE aircraft carrying AZ tail codes [left]. The Block 60 features a passive nose-mounted IR sensor, as well as Sniper target pod. It is also characterised by the snug-fitting conformal fuel tanks, enlarged dorsal spine and a number of airframe-mounted antennae for the advanced electronic warfare suite [below]. *(USAF)*

Chapter Four

The Block 40 to 52 F-16C/D at war

In January 1991, the Block 40 experienced its baptism of fire when a coalition of forces engaged Iraq's war machine and drove it out of Kuwait. In the two decades that have followed, Block 40 and Block 50 Vipers have formed the mainstay of US air power in conflicts around the globe.

OPPOSITE A Block 42 F-16 unleashes an AGM-65 Maverick. These missiles are used by all Blocks of Viper against armoured and reinforced targets. For the Block 50/52 Wild Weasels, the missile also acted as a sort of poor man's substitute for a target pod. *(USAF)*

ABOVE Munitions crews load a cluster bomb on the right inboard wing station of a Block 30F during Operation Desert Storm. This aircraft belongs to the 401st Tactical Fighter Wing, Torrejon Air Base, Spain. *(USAF)*

ABOVE Whilst capable of precision deliveries using guided munitions, the F-16 would be relegated to spend most of the war as a rather lightweight bomb truck. The yellow circles on the noses of Mk84 2,000-pound bombs – a staple munition for the F-16 during the war – denote that they are live weapons. *(USAF)*

BELOW Taking fuel from a KC-135, Torrejon's Vipers flew the first daylight strike of the war. Two are seen here, each carrying six Mk82 500lb bombs. *(USAF)*

BELOW RIGHT For self-defence, F-16s during ODS often carried four AIM-9M short-range IR missiles. Two were mounted on each wingtip, and two (as seen here) were located on the outboard wing pylons. *(USAF)*

Freeing Kuwait

The F-16C/D Block 40's combat debut came in Operation Desert Storm (ODS) in January 1991. Known also as the First Gulf War, or Gulf War I, ODS saw a coalition of forces, led by the United States, oust Iraqi forces from Kuwait.

Rapid deployment to the Arabian Peninsula from bases on the east cost of the United States required 16 hours of non-stop flying by the aircraft. A range of F-16 Blocks were deployed to the Middle East in response to Saddam Hussein's invasion of Kuwait in August 1990, from the GPU-5 cannon-armed F-16As of the New York ANG to the most modern F-16C/D Block 40/42s.

The most numerous fighter in the USAF inventory then, and now (2013), F-16s had been deployed en masse to the region during Operation Desert Shield in late 1990. With 249 examples present by the time the war began, it would also become the most numerous single type in theatre.

The Viper was employed exclusively in the air-to-ground role during ODS, meaning that it was doing exactly what had been envisioned so early on in its life – providing a strike capability in a far-off land. However, it retained a limited offensive air-to-air capability through the wingtip carriage of AIM-9 Sidewinder missiles.

The combat debut for the USAF Viper pilots predictably resulted in a number of invaluable lessons. Foremost amongst these was that the traditional dive-toss, level and dive deliveries of unguided air-to-ground munitions was simply

no longer accurate enough by comparison to newer, more high-tech methods. Brand new F-15E (with LANTIRN installed), and even the venerable General Dynamics F-111F (equipped with the Pave Tack laser designator and FLIR pod), were out in the desert 'plinking' tanks, hardened aircraft shelters, runways, taxiway intersections and all manner of other targets with laser-guided bombs. But the Viper was lobbing 'dumb bombs' in a manner that fundamentally differed little from the days of World War Two.

In reality, the problem lay not with the F-16 or its pilots. It was actually a supply and logistical problem – LANTIRN was so new that there simply weren't enough pods to go around. The vast majority of LANTIRN target pods that were in theatre had been earmarked for the F-15E (but even amongst the Strike Eagle, the AAQ-14 pods were heavily rationed and it was not unusual for only one in four F-15Es to carry it). To complicate matters further, when the war started on 17 January 1991 there were only 72 Block 40/42 F-16s in theatre that were actually wired to carry LANTIRN. Ultimately the sophisticated Block 40 reverted to the role of a simple, lightweight bomb truck that would deliver two 2,000lb Mk84 LDGP (low-drag general purpose) bombs, either visually or against a set of INS coordinates.

The Viper could employ these massive bombs accurately from a low-level dive attack, but dictator Saddam Hussein had built around him a 'Super MEZ' – a missile engagement zone supplemented by a dense thicket of low-level anti-aircraft artillery (AAA). This forced the Viper pilots to drop their bombs from greater slant ranges (the range measured from the aircraft to the target), since they had to operate at between 17–20,000ft to avoid the heaviest of the AAA.

Higher altitude deliveries meant that winds aloft reduced the accuracy of a weapon that was already marginal. The Mk82 had a circular error probability (CEP) of 30ft (a 30ft CEP means that the average distance from the target that the bomb will fall is 30ft) at low level, but from a medium altitude delivery its CEP ballooned to 200ft.

Lessons were also learned from the losses the F-16 experienced. Three Vipers were downed during the conflict – one to an SA-6, one to an SA-3, and one to AAA. Although this represented a very low loss rate from a statistical point of view, two of these Vipers were downed in one sortie on 19 January. The mission saw a very large 'Gorilla Package' reminiscent of the Vietnam Air War sent on the first daylight raid across southern Iraq and into the Baghdad Super MEZ.

This package of 64 F-16s, accompanied by Wild Weasel F-4Gs, EF-111 jammers and F-15 escorts, were soon forced to jink wildly to avoid the heavy radar-guided flak that started only 50nm into Iraq from the southern border. The jinking stretched the formation out, staggering the different parts of the package and resulting in the first Vipers arriving over the target minutes before those at the rear of the package. Meanwhile, this elastic-band effect was further exaggerated because the F110-equipped jets cruised slightly faster than those with the F100.

The EF-111s successfully jammed the threat radars in the target area, while the F-4Gs unleashed HARMs at those that came on air, so the first F-16s egressed unscathed. Infuriatingly, they also did so still laden with

ABOVE An F-16D Block 40 sits on the ramp carrying an AAQ-13 LANTIRN navigation pod and a load of Mk82 dumb bombs. During ODS, severe shortages of the AAQ-14 target pod meant that even the newest Vipers had to revert to old-fashioned bombing methods. *(USAF)*

bombs, having arrived to find the target socked in by a solid undercast. But the final elements of the package arrived overhead to find the skies clearing and the Iraqi gunners now very capable of seeing and targeting them. In the pandemonium that followed, two Vipers were downed by SA-3 and SA-6 SAMs, their pilots ejecting successfully, but destined to spend the rest of the war as prisoners.

F-16s were also used as medium-altitude bombers in an attack on the Osirak nuclear reactor (the same reactor that Israeli F-16As had struck in June 1981). Once again, the strike saw a large package (60 aircraft, of which 32 were F-16s) sent against a heavily defended target. And, once again, the attack was thwarted by poor visibility of the target area – this time caused not by meteorological phenomena, but by man-made smoke pots located around the facility. In a testament to the dominance of stealth, the facility was damaged significantly the following night by eight Lockheed F-117 Nighthawks.

The large strike packages in which the F-16 had been embedded typically targeted static targets, and success was often mixed. However, in the close air support (CAS) and battlefield air interdiction (BAI) roles, the Viper was generating a much more consistent set of positive results. Under the direction of forward air controllers, the Vipers would work in units of two or four ships, each element or section assigned to patrol a 15nm square map-grid area known as a 'kill box'.

Working these boxes back and forth, Vipers

attacked Iraqi forces with CBU-87 cluster munitions, AGM-65 Maverick missiles and 500lb and 2,000lb LDGPs. In doing so they accounted for more than 360 armoured vehicle kills. Meanwhile, the Block 40s used their GPS to help locate Scud missile launchers by night – a mission fraught with difficulties, as already experienced by the LANTIRN-equipped Strike Eagles performing the same role, and yielding limited concrete results.

By the end of the short war, the F-16 had flown 43% of all USAF strike missions, amounting for 13,480 sorties, of which around 4,000 were flown at night. The average sortie duration was 3.24 hours.

With Iraq's military pushed back to Baghdad, Operation Desert Storm came to a close. However, measured combat operations over Iraq would actually continue for more than another decade to enforce UN-approved No-Fly Zones (NFZs) over northern Iraq (Operation Northern Watch – ONW) and southern Iraq (Operation Southern Watch – OSW).

ONW was implemented first in a bid to protect the Kurds in northern Iraq from further aggression by Hussein (an irony not lost on coalition pilots, some of whom expressed dismay that Turkish aircraft could bomb them with impunity), and it provided the USAF with its first opportunity to claim a MiG kill at the hands of the F-16. On 27 December 1992 an F-16D Block 42 of the 23rd Fighter Squadron, 52nd FW, stationed at Incirlik AB, Turkey, shot down an Iraqi MiG-23 using an AIM-120 AMRAAM.

From the Desert to the Balkans

While the F-16 Block 40/42 continued to form a central component of ONW and OSW (and the Block 50 would soon join it), these advanced Vipers were very soon to be called upon to also contribute to NATO's efforts in the Balkans region.

The first of these efforts, Operation Deny Flight (ODF), came in April 1993 and saw NATO implement a No-Fly Zone over Bosnia-Herzegovina in response to UN Security Council Resolution 781. ODF also provided close air support for UN troops on the ground in Bosnia. The troops were tasked with halting the

Bosnian-Serb persecution of ethnic minorities, but were themselves under threat of air attack.

Eleven NATO nations committed their armed forces to the operation, and the Block 40 F-16CGs of Aviano's 510th and 555th Fighter Squadrons, 31st FW, gave the coalition a valuable all-weather multi-role capability, as would very soon be demonstrated.

On 28 February 1994 six Serbian Air Force J-21 Jastreb jets attacked a Bosnian factory, but as they egressed the target a pair of Aviano Vipers were waiting for them. In the melee that ensued, four Jastrebs were downed with AIM-9 Sidewinders fired from well within visual range.

ODF's air campaign was expanded in August and September 1995, when NATO switched emphasis to targeting Serb air defence systems. The change, code-named 'Dead Eye', heralded the first sustained F-16CJ Wild Weasels combat operations. It was named Operation Deliberate Force.

The CJs came from the 52nd Wing, Spangdahlem AB, Germany. Whilst Aviano's Block 40s would also participate in pre-emptive strikes during Deliberate Force (and would lose one Viper to an SA-6 surface-to-air missile), it was the Wild Weasels that saw most of the action.

With tensions in the region worsening, Operation Noble Anvil (ONA) followed Deliberate Force. In fact, ONA is more accurately the name given by US commanders to America's role in Operation Allied Force (OAF), from 24 March to 11 June 1999. Once again a NATO operation, OAF came in response to ethnic cleansing by the Federal Republic of Yugoslavia under the

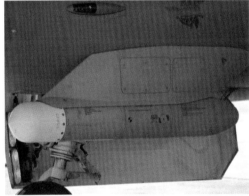

leadership of Slobodan Milošević. Milošević's forces were being challenged by the Kosovo Liberation Army, which also sought independence for Kosovo. To settle the issue, the Rambouillet peace accords were signed by Albania, the United States and Great Britain, calling for NATO to commit 30,000 peacekeepers to Kosovo and to enforce right of free passage through Yugoslav territory. Milošević refused to sign. NATO prepared for conflict.

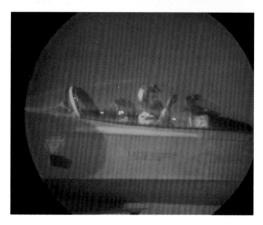

The Viper once again figured prominently in the operations that were to follow. The United States assigned 64 F-16CG and F-16CJ Vipers from Aviano and Spangdahlem, the latter's Wild Weasels joining the former's Night Falcons at Aviano to become the 31st Air Expeditionary Wing. In fact, a multinational force of Vipers was assembled, with America's F-16s joined by earlier Blocks from Portugal, the Netherlands, Norway and Belgium. The entire F-16 force was tasked with the destruction of Serbian infrastructure, as well as those armed forces that continued to perpetrate criminal acts against civilians.

Among the first to enter Serbian airspace on 24 March were four F-16AMs of 322 Sqn, Royal Netherlands Air Force, based at Ameldola Air Base, Italy. The action that followed was intense. One of the four Dutch Vipers downed a MiG-29 Fulcrum with an AIM-120 AMRAAM, while two F-15Cs of the 393rd FS, 48th FW, RAF Lakenheath, England, downed two more Fulcrums within a matter of minutes. With this going on, the 22nd and 23rd FSs from Spangdahlem employed their AGM-88 HARMs against radar sites, and the 510th and 555th FSs from Aviano assisted using freefall weapons. So intense was the effort that the majority of the Silver Star gallantry medals awarded during OAF went to F-16 pilots who had taken part in the hunting and killing of Serb SA-2, SA-3 and SA-6 SAM sites.

Even though the mobile SA-6s proved notoriously difficult for F-16CJs to find and neutralise, the fixed SAM sites and AAA were also menacing. The 555th FS lost a jet to an

SA-3 over Serbia on 2 May, and his wingman was later awarded the Silver Star for marshalling rescue forces to him in the face of enemy fire. On 4 May the 52nd FW's Vipers finally got in on the air-to-air action, when a 22nd FS Viper downed a lone FRY MiG-29 near Belgrade. The Balkan conflicts had given the F-16 the opportunity to truly exploit its multi-role pedigree and establish the type as an excellent fighter.

Large-scale NFZ operations

While the Vipers of Aviano and Spangdahlem were kept busy in the Balkans, US Air Force Vipers from around the globe continued to participate in the No-Fly Zone policing operations over Iraq. Indeed, the F-16 had maintained a constant presence in the region since August 1990 by the time a set of concentrated, coordinated strikes on Saddam Hussein's weapons-making and storage facilities was launched in 1998. It was called Operation Desert Fox.

Desert Fox commenced with cruise missile attacks against military targets in Iraq, just as Operation Desert Storm had in January 1991. This operation, which effectively 'interrupted' Northern and Southern Watch, responded to Iraq's continued failure to comply with United Nations Security Council resolutions on nuclear, chemical and biological weapons inspections.

Designed to damage Hussein's capability to make or deliver weapons of mass destruction, F-16s of the 4404th (Provisional) Wing based at both Prince Sultan Air Base, Saudi Arabia,

ABOVE LEFT AND ABOVE F-16CGs from Hill AFB and F-16CJs from Shaw AFB worked in close partnership during Operation Desert Fox. Along with other coalition assets, they struck strategic targets in December 1998. The bird slicer antennae and the PACER GEM paint indicate that these images actually post-date that operation. *(USAF)*

and Ahmed Al Jaber Air Base, Kuwait, took part. These comprised 16 F-16CJ Wild Weasels of the 20th Fighter Wing, Shaw AFB, SC; and 20+ F-16C/D Block 40s from the 388th Fighter Wing, Hill AFB, Utah. On 16 December the F-16s took part in mass actions that struck 50 separate targets. On 19 December President Clinton announced a cessation to the campaign and the F-16s resumed their long vigil over Iraq.

Operation Enduring Freedom

In October 2001, in the wake of the 9/11 terrorist attacks, the United States led a coalition force into Afghanistan for Operation

BELOW A 555th Expeditionary Fighter Squadron F-16CM Block 40 receives fuel from a KC-10 Extender over Afghanistan during Operation Enduring Freedom. The 555th EFS is the expeditionary forces' designation for the 'Triple Nickel', ordinarily stationed at Aviano AB, Italy, with sister unit, the 510th FS. The Viper is loaded with four GBU-38 500lb JDAMs. *(USAF)*

ABOVE Loaded for business, an F-16CJ (now CM) flashes a pair of AGM-88 HARMs. The missile would account for a good number of initial radar kills during Operation Iraqi Freedom. (USAF)

BELOW A pilot calls a redball prior to a combat sortie over Iraq in 2003. The expert skills of the F-16 maintainers ensured that the Viper sortie ratio stayed high throughout the conflict. (USAF)

Enduring Freedom. Meanwhile, in the continental United States, Operation Noble Eagle (ONE) was launched. This involved F-16s armed with AIM-120 and AIM-9 missiles patrolling major cities and infrastructure sites to ensure that any further hijacked airliners were shot down before they could reach their targets. ONE continues in 2013 and is a crucial element of homeland defence.

For Enduring Freedom, USAF F-16s targeted Taliban infrastructure and personnel in Afghanistan, but their European colleagues soon joined them. Based at Manas AB, Kyrgyzstan, Norwegian, Dutch and Danish F-16s flew operations from October 2002 to October 2003, although Norway's aircraft returned home early.

The Royal Netherlands Air Force returned to Manas AB with six F-16s on 10 September 2004, providing assistance to the NATO-led ISAF (International Security Assistance Force) during the Afghan presidential elections. The aircraft

returned to the Netherlands on 19 November 2004. On 14 July 2005 they returned once more, this time accompanied by four Belgian Air Force F-16s – the first deployment by the Belgians – thus forming the DATF (Deployable Air Task Force). Once more, the deployment was in support of the ISAF mission to provide additional security during the Afghan elections.

OEF continues into July 2013, and F-16s form a central pillar of the air support element of the operation.

Shock and awe: the Second Gulf War

On the evening of 19 March 2003, the United States commenced Operation Iraqi Freedom – the Second Gulf War (Gulf War II). The first priority of OIF was to take control of the battle space, necessitating an extensive SEAD effort to remove key individual components of Iraq's integrated air defence system (IADS). This would require all 71 F-16CJs to be deployed to the theatre of operations, as well as the might of the US Navy's EA-6B Prowlers and F/A-18 Hornets.

The USAF commitment to this objective was spearheaded by the F-16CJs of the 77th FS, 20th FW, Shaw AFB, SC. The 'Gamblers' were deployed to Prince Sultan Air Base (PSAB), Saudi Arabia, where they would be the dominant squadron in the 363rd Air Expeditionary Group (AEG). The 363rd AEG would also be home to eight Block 50 F-16CJs from the 14th FS, 35th FW, Misawa AB, Japan; six Block 40 F-16CGs from the 4th FS, 388th FW, Hill AFB; and six Block 30 F-16Cs from the

457th FS, 301st FW, USAF Reserve. Combined, these F-16s were tasked with defence of the battle space and other aircraft.

The 524th FS, 27th FW, Cannon AFB, was tasked to provide 18 of its Block 40 F-16CGs to Al Jaber AB, Kuwait, as part of the 332nd AEG. Like the 14th FS in Saudi Arabia, the 524th FS 'Hounds' had deployed into the theatre in December 2002 as part of the Air Expeditionary Forces (AEF) structure that was used to enforce the No-Fly Zones. Twelve jets initially deployed, but another six arrived later when the squadron 'plussed-up' for seemingly inevitable hostilities. These Vipers were to provide alert and interdiction capabilities, and were part of the air defence plan to protect Kuwait. They would scour the border between Iraq and Kuwait, providing non-traditional information surveillance and reconnaissance (NTISR) intelligence on border activities or incursions. The 524th FS was also slated to serve as a key platform to drop M129 leaflet dispensers as part of an expansive psychological warfare effort.

Perhaps the most important mission – politically, at least – was assigned to the men and women of the 410th Air Expeditionary Wing (AEW), which comprised the Alabama Air National Guard (ANG) as the lead wing, plus Vipers from the Colorado and Washington DC ANGs, and a contingent of Air Force Reserve (AFRES) F-16s from 466th FS, 419th FW, Hill AFB. The 190th FS, 187th FW, AL ANG had some of the most capable F-16s in the Air Force, although these were Block 30/32 airframes. Key to their potency were a number of recent OFP upgrades and the introduction of the new Situational Awareness Data Link.

Colorado's 120th FS, 140th Wing, and Washington's 121st FS, 113th FW, Block 30 Vipers would help the Alabama F-16Cs hunt for Scud missiles in what was simply called the 'anti-TBM' (theatre ballistic missile) mission.

The fourth and final F-16 deployment had been planned to bed down at Incirlik AB, Turkey, in order to force Iraq to deal with a multi-front conflict. However, last-minute negotiations between Turkish officials and the US State Department failed to yield an

RIGHT Dust storms would arrive at any time, but only the worst could stop the F-16 from flying. Note the use of reflective covers in the inside of the canopy – an attempt to reduce cockpit temperatures in the blistering midday sun. *(USAF)*

agreement so these forces were relocated to the southern front. F-16CJs of the 22nd FS and 23rd FS, 52nd FW, Spangdahlem AB, Germany, were duly reassigned to Al Udeid AB, Qatar. The 379th AEW, to which the Spang' CJs were assigned, was also home to F-16CJs of the 157th FS, 169th FW, South Carolina ANG (SCANG).

The three CJ squadrons were assigned SEAD (suppression of enemy air defences), DEAD (destruction of enemy air defences), time sensitive tasking (TST) and close air support (CAS) as their primary missions.

Opening gambits

19 March saw the phased changeover from OSW to the enigmatically named OPLAN 1003V – the combat plan for OIF. The move to OPLAN 1003V had been brought forward by two days in order to take advantage of time sensitive intelligence, thus catching many coalition aircrew by surprise. One F-16CJ pilot flying an OSW sortie was caught completely unawares when he was re-tasked in flight to provide support for an F-117 sent to strike a site purported to be housing Saddam Hussein. In the event, Hussein was not there. But the F-16 had just been part of the opening gambit of the war. OIF had begun.

In the early stages of the war, the Wild Weasel Viper was often assigned to the SEAD Escort role, each jet loaded with two HARMs and assigned to protect a coalition package of strikers. SEAD Escort required the F-16CJs to fly in close coordination with the strikers on their way to the target, sniffing out radar nodes and threats, and pre-emptively employing the AGM-

88 from stand-off ranges in a bid to keep threat radars offline.

The AGM-88's pre-planned mode made this an effective mission – the HARM was fired at a suspected site regardless of whether it was emitting. Based on precise and thorough pre-flight planning, the timings were calculated so that the HARM would be in the air as the strikers were over the target – their most vulnerable point. If the threat radar came online during the missile's time of flight, the HARM's sensitive seeker would detect the radar energy and issue corresponding guidance commands to steer the missile towards the source. The main priority was to get threat emitters off-air as soon as possible, or to dissuade them from coming online at all.

In keeping with its role to defend the battle space and other aircraft, the 77th EFS and 14th EFS were concentrated on dealing with the SAM threat around Baghdad. The Super MEZ comprised over 200 SAM systems, many of which were never pinpointed accurately, and it spilled over into the cities surrounding Baghdad. Whilst the 22nd EFS and 23rd EFS at Al Udeid would target the very heart of the MEZ, the 'Gamblers' (77th EFS) and 'Samurais' (14th EFS) would pick off elements of the SAM structure around Baghdad as coalition strike requirements dictated. The effort resulted in dozens of SAM launches at the F-16s – some radar guided, some optically guided and others unguided – and in the ensuing engagements the Wild Weasel pilots certainly earned their reputation for performing the most dangerous mission of all.

Just as it had in 1991, the battle space across Iraq had been split up into kill boxes (this time measuring 10nm x 10nm) to allow strikers assigned to CAS and BAI to work specific areas and to provide de-confliction with other coalition air assets. But the F-16CJs were allowed to roam wherever they wished during their on-call SEAD vulnerability times.

When tasked to fly support in one of the kill boxes, there was no telling exactly what they'd be called to do. Sometimes they struck hard targets such as ammunition bunkers, other times Special Forces (SF) troops would call that they were under fire and needed support. Occasionally a show of force was required. On several other occasions the F-16CJ was called in to use its 20mm M61-A1 Gatling gun to strafe soft mobile targets on the ground.

Within days of the start of OIF it became readily apparent that the Iraqi IADS was not going to fight to its fullest capacity. More often than not, the F-16CJ's HTS detected only spurious transmissions, and the most potent threat came from mobile systems. The requirement to change tactics and mission type became increasingly apparent as more and more Wild Weasels returned home from sorties with a full weapons load on board.

Accordingly, the F-16CJs of the 52nd Wing and SCANG rolled into a precision strike DEAD role: they employed JDAM (joint direct attack munitions) and WCMD (wind-corrected munitions dispensers) against visually-acquired

threats, pre-planned targets, and in support of forward air controllers. Pre-planned targets were often fixed SAM sites – on one sortie an F-16CJ dropped a JDAM on an SA-2 launcher and then rolled in and released a WCMD on an adjacent SA-3 site while the JDAM was still in the air. On these missions pairs of Wild Weasels flew with mixed loads: one jet carried HARMs, and the other GPS-guided weapons (typically a JDAM on one wing and WCMD on the other).

Incredibly, history repeated itself to some degree once again. Whilst the AAQ-14 LANTIRN target pod had by this time been superseded by the AAQ-28 Litening II target pod, there were not enough Litenings to go around, and the F-16CG and Block 30 squadrons had the bulk of those that were in

ABOVE As the sun goes down, an F-16CG pilot climbs up to the cockpit. Operations over Iraq went on night and day. They were relentless. *(USAF)*

LEFT Armed with the most basic of load-outs – four Mk82 LDGP bombs – an F-16C Block 30 from Hill AFB's 466th FS 'Diamondbacks' sits ready for the next sortie over Iraq. *(USAF)*

theatre. The CJ's inability to find and designate targets from altitudes that kept it away from lethal man-portable air defence systems (MANPADS, which home in on the target's exhaust heat signature) was an inconvenience. F-16CJ pilots were dependent upon assistance for much of the war to find, positively identify (PID) and conduct collateral damage estimates (CDE) on their targets. Daytime operations saw exceptions to the rule, but it would seem that for those who flew at night, the addition of a target pod would have made a great difference.

With organised opposition thinning, the F-16CJs began flying both DEAD and flexible tasking sorties, armed with two AGM-65 Mavericks and supported by a wingman with two JDAMs. The AGM-65 was a useful weapon not only because of its excellent weapons effects, but also because with a shortage of target pods it was the closest they came to using an electro optical or IR sensor system to find and attack targets.

At Ahmed Al Jaber AB, Kuwait, the 524th EFS was busy using its Block 40s to great effect, having already expended 176,500lb of munitions in OSW strikes before OIF had even begun. These were mainly some 87 2,000lb JDAMs that the squadron, taking advantage of Iraqi threat reactions as an opportunity to prosecute the OIF 'battlefield preparation' phase, employed against Iraqi

cable repeater sites, thus dismembering the enemy's communications network. Additionally, five GBU-12s had been used against Iraqi air defence assets.

With OIF combat operations now under way, the squadron was assigned to destroy, amongst other things, communication sites just west of Baghdad. The targets were small buildings about 50 miles west of Baghdad that housed fibre optic switching equipment, and they were struck with BLU-109s – the penetrator version of the GBU-31 2,000lb JDAM.

As an example of the sort of combat activity experienced by an F-16CJ squadron, the 77th EFS employed 170 CBU-103, 105 AGM-88, 52 GBU-31, 16 AGM-65 and in excess of 7,000 rounds of 20mm PGU-28. The squadron engaged 338 ground targets, destroyed 104 SAMs, radars and AAA pieces, and also destroyed or disabled 20 tanks and APCs, 26 trucks and 36 aircraft. Shaw's pilots and maintainers had flown/generated 676 sorties for a total of 3,803.5 combat flying hours by the time the 'shooting war' ended.

Libya

By 2009 the situation in Iraq had stabilised dramatically and the United States and Great Britain had withdrawn the majority of their

RIGHT Four GBU-12s hang beneath the wings of a Viper on patrol over Iraq. Seen here dropping flares for the camera, the image is dated 2004. By this time the 'shooting war' was long over and the F-16 was increasingly being used to strike insurgents in urban settings. (USAF)

combat forces, leaving behind only 'advisors'. This scaling down was reflected by the end of OIF and the start, in February 2010, of Operation New Dawn.

However, in North Africa new troubles were brewing. The people of Libya had risen against their dictator, Colonel Muammar Gaddafi, and late in 2010 this 'Arab Spring' gathered pace. In response the UN Security Council passed resolutions 1970 and 1973 in February and March 2011, allowing for a range of sanctions against Gaddafi. From a military perspective, the resolutions permitted the creation of a No-Fly Zone and the use of force against pro-Gaddafi forces. NATO enacted these powers under Operation Unified Protector.

OUP saw the 52nd FW's F-16CM Wild Weasels once again co-locate with the 31st FW's F-16CM Block 40s at Aviano AB, Italy. Aviano would also play host to a range of international Vipers. However, President Obama had been clear that the United States would not lead the military operation to help 'liberate' Libya. It was a mix of international F-16s of various Blocks – Belgian, Danish and Norwegian F-16AM Block 15 MLUs; Jordanian F-16A/B Block 20 MLUs; UAE Block 60 Desert Falcons; and Greek F-16C/D Block 52+s – that picked up the majority share of the Viper's sorties over Libya.

For Greece, Unified Protector marked the

combat debut of their Block 52+ aircraft, notwithstanding the frequent skirmishes with Turkish fighters that occur over the Aegean Sea. Likewise, it also signalled the F-16E's debut in combat, with the UAE Vipers flying out of Decimomannu, Italy, and providing air cover for other coalition aircraft.

Norway's F-16s ceased operations at the end of July 2012, at which point they had accounted for 10% of the total sorties by the coalition and had dropped 600 munitions against targets that included armour, installations and ammunition dumps. Meanwhile, Belgian Vipers flew 448 missions, dropping 365 bombs with a success rate of 97%.

ABOVE A 77th Fighter Squadron pilot goes through his checks prior to taxi and take-off. The extensive mission markings on the side of the aircraft show that it has had a busy combat deployment to Iraq or Afghanistan. *(USAF)*

LEFT An F-16CM Block 50 from the 13th FS slides towards the tanker during an Operation New Dawn sortie in April 2011. The aircraft is equipped with both Sniper and HTS pods on its chin hardpoints. *(USAF)*

Chapter Five

Anatomy of the Viper F-16CM

Blocks 40 to 52

The technical specifications provided in this chapter and the following are taken from a range of official technical texts published by the US Air Force and available in the public domain. Where appropriate, additional information has been added for the sake of clarity, and the author has reworded some passages for the same reason. However, in essence the following description very closely resembles the F-16CM technical orders that US Air Force pilots and maintainers are expected to read and memorise.

OPPOSITE With the radome and panels open, the Viper's electronic innards are revealed. Light grey boxes - also known as line replaceable units (LRUs) - can be pulled and replaced by maintainers on the flightline. *(USAF)*

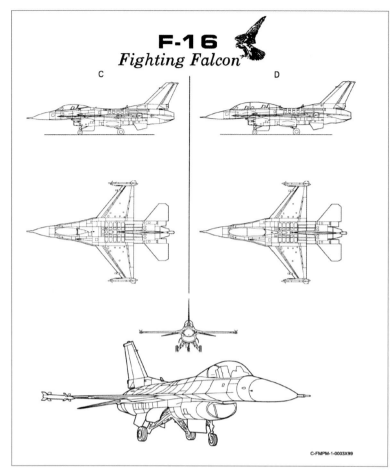

F-16
Fighting Falcon

C   D

C-FMPM-1-0003X99

General overview

The F-16CM fuselage is characterised by a large bubble canopy, wingbody strakes, and an under-fuselage engine air inlet. The wing and tail surfaces have a moderate aft sweep. The wing has automatic leading-edge flaps and trailing-edge flaperons that function as both flaps and ailerons. Wingspan, including tip-mounted AIM-120 AMRAAM missiles, is 32ft 10in.

The cockpit arrangement is conventional, although the ACES II ejection seat is reclined 30°. The control stick is mounted on the right console.

The horizontal tails have a negative dihedral and provide pitch and roll control through deflections that can be either symmetric or differential. The vertical tail provides directional stability. It is augmented by twin ventral fins mounted on the underside of the aft fuselage.

All of the F-16CM's flight control surfaces are actuated hydraulically, the two independent hydraulic systems being controlled by signals through a fly-by-wire flight control system.

Maximum operating airspeed is 800kt from sea level to 30,000ft MSL (mean sea level).

WINGS
Area	300 Sq Ft
Span	30 Ft
Aspect Ratio	3.0
Taper Ratio	0.2275
Sweep (LE)	40°
Dihedral	0°
Airfoil	NACA 64A204
Incidence	0°
Twist	
At BL 54.0	0°
At BL 180.0	3°
Flaperon Area	31.32 Sq Ft
LEF Area	36.71 Sq Ft

HORIZONTAL TAILS
Area	63.70 Sq Ft
Aspect Ratio	2.114
Taper Ratio	0.390 (Theo)
Sweep (LE)	40°
Dihedral	-10°
Airfoil	
At Root	6% Biconvex
At Tip	3.5% Biconvex

VERTICAL TAIL
Area	54.75 Sq Ft
Aspect Ratio	1.294
Taper Ratio	0.437
Sweep (LE)	47.5°
Airfoil	
At Root	5.3% Biconvex
At Tip	3.0% Biconvex
Rudder Area	11.65 Sq Ft

SPEEDBRAKES
Area (4 Element Clamshell)	14.26 Sq Ft (3.565 Sq Ft Ea)

VENTRAL FIN (EACH)
Area	8.03 Sq Ft
Span	23.356 In. Theo (27.5 In. Actual)
Aspect Ratio	0.472 (Theo)
Taper Ratio	0.760 (Theo)
Sweep (LE)	30°
Dihedral (Cant)	15° Outboard
Airfoil	
At Root	3.886% Modified Wedge
At Tip	Constant 0.03R

LANDING GEAR (LG)
Main Gear (MLG)	
Tire Size	27.75 x 8.75-14.5 24 Ply
Stroke	10.5 In.
Static Rolling Radius	11.0 In.
Nose Gear (NLG)	
Tire Size	18 x 5.7-8 18 Ply
Stroke	10.0 In.
Static Rolling Radius	7.5 In.

ENGINE
F100-PW-220/220E	
Thrust	25,000 Lb Class
Compressor Diameter	34.8 In.
Engine Length	191.1 In.
F100-PW-229	
Thrust	29,000 Lb Class
Compressor Diameter	34.8 In.
Engine Length	208 In.
F110-GE-100	
Thrust	28,000 Lb Class
Compressor Diameter	35.8 In.
Engine Length	183.76 In.
F110-GE-129	
Thrust	29,500 Lb Class
Compressor Diameter	35.8 In.
Engine Length	183.76 In.

32 FT 10 IN. *
OVERALL SPAN W/MISSILES
31 FT
OVERALL SPAN W/O MISSILES
10°
46.5
93.0
* Add 3 inches for AIM-120 missiles
CM-FMPM-1-0006X99

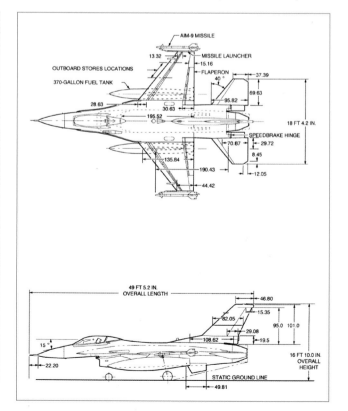

AIM-9 MISSILE
MISSILE LAUNCHER
OUTBOARD STORES LOCATIONS
FLAPERON
370-GALLON FUEL TANK
13.32
15.16
37.39
40°
69.63
95.82
28.63
30.63
195.52
18 FT 4.2 IN.
SPEEDBRAKE HINGE
70.87
29.72
135.84
8.45
190.43
12.05
44.42

49 FT 5.2 IN.
OVERALL LENGTH
46.80
15.35
82.05
95.0 101.0
29.08
15°
108.62
19.5
22.20
16 FT 10.0 IN.
OVERALL HEIGHT
STATIC GROUND LINE
49.81

Above 30,000ft MSL the aircraft is limited to 2.05 Mach.

Length: 49ft 5.2in.
Height (to top of vertical tail): 16ft 10in.
Height (to top of canopy): 9ft 4in.
Tread: 7ft 9in.
Wheelbase: 13ft 2in.

Gross weights

Gross weight varies with engine installation choice. Following the completion of the CCIP programme, F-16C/Ds in Blocks 40 through 52 weigh approximately the same. The following approximate data includes pilot, oil, two AIM-120 AMRAAM missiles, a full load of 20mm ammunition, full CCIP upgrades and internally carried JP-8 fuel:

F-16C, PW220: 26,900lb (19,700lb JP-8).
F-16D, PW220: 26,300lb (19,700lb JP-8).
F-16C, PW229: 27,500lb (20,300lb JP-8).
F-16D, PW229: 26,900lb (21,000lb JP-8).
F-16C, GE100: 27,700lb (20,500lb JP-8).
F-16D, GE100: 27,100lb (21,200lb JP-8).
F-16C, GE129: 27,800lb (20,600lb JP-8).
F-16D, GE129: 27,200lb (21,300lb JP-8).

Electrical system

The F-16CM's electrical system consists of a main ac (alternating current) power system, standby ac power system, emergency ac power system, dc (direct current) power system, FLCS (flight control system) power supply, and provisions for external ac power.

The ac power is normally supplied by a 60kva main generator located on, and driven by, the accessory drive gearbox (ADG, attached to the engine). This generator supplies power to panels that protect against overcurrent, and to three buses: nonessential, essential, and emergency ac. Eight overcurrent sensing contactors protect certain ac buses, external stores stations 3, 5, and 7, and the two chin stations from overcurrent.

The standby ac power system comprises the essential and emergency ac buses. When the main generator is off or inoperative, it is powered by a 10kva standby generator located on and driven by the ADG. The standby generator itself features an integral 'FLCS permanent magnet generator' (PMG) that supplies power to the FLCS.

If the main and standby generators fail,

BELOW The electrical distribution schematic for the Block 40/42. *(USAF)*

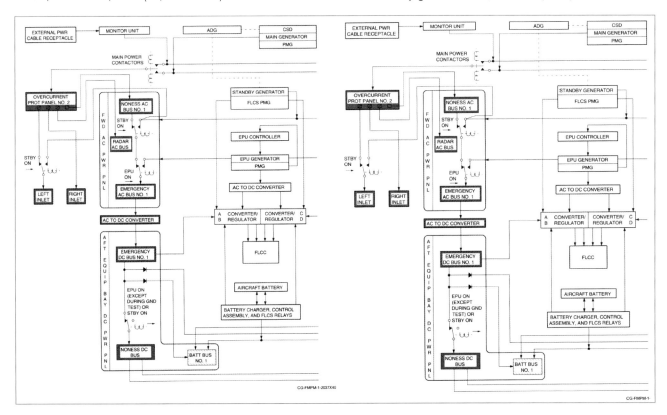

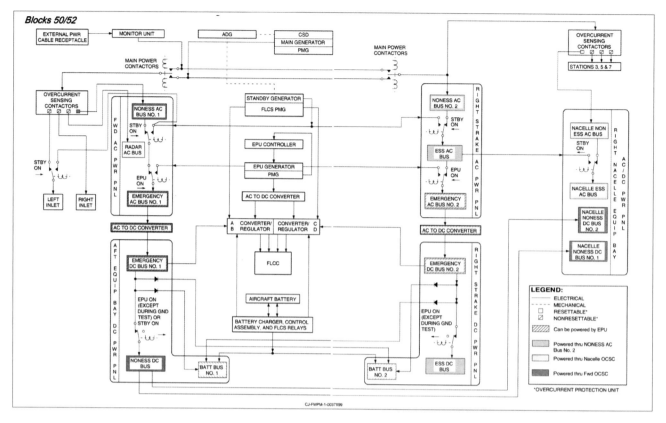

ABOVE The electrical distribution schematic for the Block 50/52. *(USAF)*

BELOW The AC electrical distribution diagram for the F-16CM shows the systems powered by the nonessential and essential buses in addition to the emergency buses. *(USAF)*

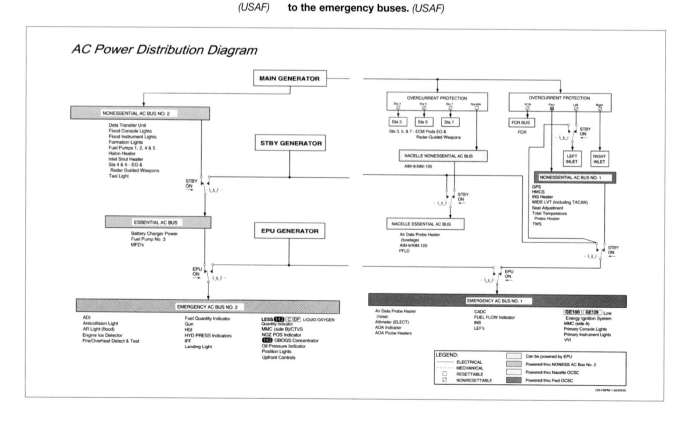

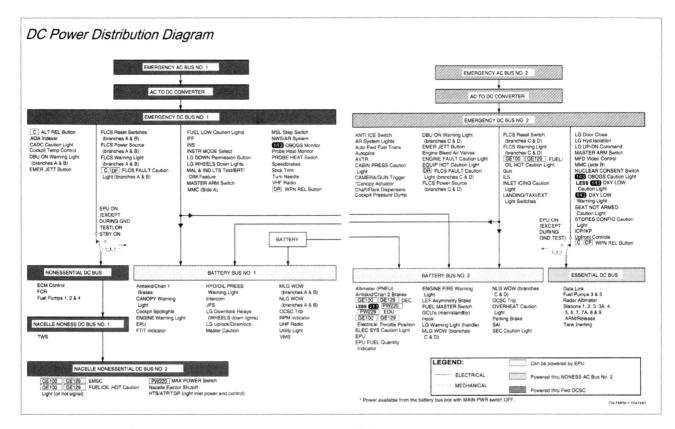

DC Power Distribution Diagram

EMERGENCY AC BUS NO. 1 → **AC TO DC CONVERTER** → **EMERGENCY DC BUS NO. 1**

EMERGENCY AC BUS NO. 2 → **AC TO DC CONVERTER** → **EMERGENCY DC BUS NO. 2**

Under EMERGENCY DC BUS NO. 1:
- [C] ALT REL Button
- AOA Indexer
- CADC Caution Light
- Cockpit Temp Control
- DBU ON Warning Light
- EMER JETT Button

- FLCS Reset Switches (branches A & B)
- FLCS Power Source (branches A & B)
- FLCS Warning Light (branches A & B)
- [C] [DF] FLCS FAULT Caution Light (branches A & B)

- FUEL LOW Caution Lights
- IFF
- INS
- INSTR MODE Select
- LG DOWN Permission Button
- LG WHEELS Down Lights
- MAL & IND LTS Test/BRT/DIM Feature
- MASTER ARM Switch
- MMC (Side A)

- MSL Step Switch
- NWS/AR System
- [143] OBOGS Monitor
- Probe Heat Monitor
- PROBE HEAT Switch
- Speedbrakes
- Stick Trim
- Turn Needle
- VHF Radio
- [DR] WPN REL Button

Under EMERGENCY DC BUS NO. 2:
- ANTI ICE Switch
- AR System Lights
- Auto Fwd Fuel Trans
- Autopilot
- AVTR
- CABIN PRESS Caution Light
- CAMERA/GUN Trigger
- *Canopy Actuator
- Chaff/Flare Dispensers
- Cockpit Pressure Dump

- DBU ON Warning Light (branches C & D)
- EMER JETT Button
- ENGINE FAULT Caution Light
- EQUIP HOT Caution Light
- [DR] FLCS FAULT Caution Light (branches C & D)
- FLCS Power Source (branches C & D)

- FLCS Reset Switch (branches C & D)
- FLCS Warning Light (branches C & D)
- [GE100] [GE129] FUEL/OIL HOT Caution Light
- Gun
- ILS
- INLET ICING Caution Light
- LANDING/TAXI/EXT Light Switches

- LG Door Close
- LG Hyd Isolation
- LG UP-DN Command
- MASTER ARM Switch
- MFD Video Control
- MMC (side B)
- NUCLEAR CONSENT Switch
- [143] OBOGS Caution Light
- LESS [143] OXY LOW Caution Light
- [143] OXY LOW Warning Light
- SEAT NOT ARMED Caution Light
- STORES CONFIG Caution Light
- EPU
- ICP/IKP
- Upfront Controls
- [C] [DF] WPN REL Button

EPU ON (EXCEPT DURING GND TEST) OR STBY ON

BATTERY

NONESSENTIAL DC BUS
- ECM Control
- FCR
- Fuel Pumps 1, 2 & 4

BATTERY BUS NO. 1
- Antiskid/Chan 1 Brakes
- CANOPY Warning Light
- Cockpit Spotlights
- ENGINE Warning Light
- EPU
- FTIT Indicator

- HYD/OIL PRESS Warning Light
- Intercom
- JFS
- LG Downlock Relays (WHEELS down lights)
- LG Uplock/Downlock

- MLG WOW (branches A & B)
- NLG WOW (branches A & B)
- OCSC Trip
- RPM Indicator
- UHF Radio
- Utility Light
- VMS

NACELLE NONESS DC BUS NO. 1
- TWS

BATTERY BUS NO. 2
- Altimeter (PNEU)
- Antiskid/Chan 2 Brakes
- [GE100] [GE129] DEC
- LESS [210] [PW220]
- [PW229] EDU
- [GE100] [GE129]
- Electrical Throttle Position
- ELEC SYS Caution Light
- EPU
- EPU FUEL Quantity Indicator

- ENGINE FIRE Warning Light
- LEF Asymmetry Brake
- FUEL MASTER Switch
- GCU's (main/standby)
- Hook
- LG Warning Light (handle)
- MLG WOW (branches C & D)

- NLG WOW (branches C & D)
- OCSC Trip
- OVERHEAT Caution Light
- Parking Brake
- SAI
- SEC Caution Light

ESSENTIAL DC BUS
- Data Link
- Fuel Pumps 3 & 5
- Radar Altimeter
- Stations 1, 2, 3, 3A, 4, 5, 6, 7, 7A, 8 & 9
- ARM/Release
- Tank Inerting

EPU ON (EXCEPT DURING GND TEST)

NACELLE NONESSENTIAL DC BUS NO. 2
- [GE100] [GE129] EMSC
- [GE100] [GE129] FUEL/OIL HOT Caution Light (oil hot signal)
- [PW220] MAX POWER Switch
- Nacelle Ejector Shutoff
- HTS/ATP/TGP (right inlet power and control)

LEGEND:
- ──── ELECTRICAL
- - - - - MECHANICAL
- [ ] Can be powered by EPU
- Powered thru NONESS AC Bus No. 2
- Powered thru Fwd OCSC

* Power available from the battery bus box with MAIN PWR switch OFF.

CM-FMPM-1-0042X40

---

emergency ac power is supplied automatically to the emergency ac buses via a 5kva EPU (emergency power unit) generator. The EPU generator also has a PMG, allowing it to supply dc power to the four FLCS branches through an ac to dc converter.

DC power is supplied by ac-to-dc converters, or by the aircraft battery. The dc convertors power different buses, depending upon which generator is in use. With the main generator operating, these power emergency dc bus No1, battery bus No1, nonessential dc bus, nacelle dc bus, emergency dc bus No2, essential dc bus, and battery bus No2. With the standby generator operating, they power emergency dc bus No1, battery bus No1, emergency dc bus No2, essential dc bus, and battery bus No2. Finally, with the EPU generator operating, they power emergency dc bus No1, battery bus No1, emergency dc bus No2, and battery bus No2.

If the main, standby or EPU generators are operating, the aircraft battery is disconnected and begins charging. If all generators fail, the battery is connected and powers battery bus No1 and battery bus No2. When needed, the battery buses also provide power to the FLCS power supply and start power to the EPU.

The primary FLCS power supply consists of the dedicated FLCS PMG and two dual-channel converter regulators, and four branch power supplies within the FLCS. Power redundancy is ensured by enabling other sources to power the FLCS: the main generator, standby generator, EPU generator, EPU PMG and the aircraft battery.

The FLCS PMG is the primary power source during normal operations and generates power whenever the ADG is rotating. The PMG has four outputs – one for each branch of the FLCS. At 40% rpm or greater it generates sufficient power to operate the FLCS.

Two converter/regulators provide a separate channel for each branch of the FLCS. They convert ac power from the FLCS PMG to dc power, select the power source with the highest allowable voltage, and provide dc power to the respective FLCC branch. Their output voltages are regulated to prevent overvoltage to the FLCS. The converter/regulators also provide fault indications for display on the pilot's ELEC control panel and provide test indications to the TEST switch panel.

ABOVE The DC electrical distribution diagram for the F-16CM. *(USAF)*

F-16C Block 40.

(Mike Badrocke)

1 Pitot head/air-data probe
2 Glass-fibre radome
3 Lightning conducting strips
4 Planar radar scanner
5 Radome hinge point, opens to starboard
6 Scanner tracking mechanism
7 ILS glideslope aerial
8 Radar mounting bulkhead
9 Incidence vane, port and starboard
10 IFF aerial
11 GBU-15 laser-guided glide bomb
12 AN/APG-68 digital pulse-doppler, multi-mode radar electronics equipment bay
13 Radar warning antennae, port and starboard
14 Front pressure bulkhead
15 Static ports
16 Fuselage forebody strake fairing
17 Forward avionics equipment bay
18 Canopy jettison charge
19 Instrument panel shroud
20 Instrument panel multifunction CRT head-down displays
21 Side-stick controller (fly-by-wire control system)
22 Video recorder
23 GEC wide-angle head-up display
24 Penguin air-to-surface anti-shipping missile (Norwegian aircraft)
25 LAU-3A 19-round rocket launcher
26 2.75in (68mm) FFAR
27 ATLIS II laser-designating and ranging pod
28 Intake flank (No5R) stores pylon adapter
29 LANTIRN (FLIR) targeting pod
30 Frameless bubble canopy
31 Ejection seat headrest
32 McDonnell-Douglas ACES II zero-zero ejection seat
33 Side console panel
34 Canopy frame fairing
35 Canopy external emergency release
36 Engine throttle lever incorporating HOTAS (hands-on-throttle-and-stick) radar controls
37 Canopy jettison handle
38 Cockpit section frame construction
39 Boundary-layer splitter plate
40 Fixed-geometry air intake
41 Nosewheel, aft retracting
42 LANTIRN (FLIR/TFR) navigation pod
43 Port intake flank (No5L) stores pylon adapter
44 Forward position light
45 Rapport III threat-warning antenna fairing (Belgian and Israeli aircraft)
46 Intake duct framing
47 Gun gas suppression muzzle aperture
48 Aft avionics equipment bay
49 Cockpit rear pressure bulkhead
50 Canopy hinge point
51 Ejection-seat rails
52 Canopy rotary actuator
53 Conditioned air outlet duct
54 Canopy sealing frame
55 Canopy aft glazing
56 600US gal (500Imp gal/2,271-litre) long-range fuel tank
57 Garrett turbine emergency power unit (EPU)
58 EPU hydrazine fuel tank
59 Fuel tank bay access panel
60 Forward fuselage bag-type fuel tank. Total internal capacity 6,972lb (3,162kg)
61 Fuselage upper longeron
62 Conditioned air ducting
63 Cannon barrels
64 Forebody frame construction
65 Air system ground connection
66 Ventral air conditioning system equipment bay
67 Centreline 300US gal (250Imp gal/1,135-litre) fuel tank
68 Mainwheel door hydraulic actuator
69 Mainwheel door
70 Hydraulic system ground connectors
71 Gun bay ventral gas vent
72 GE M61A1 20mm rotary cannon
73 Ammunition feed chute
74 Hydraulic gun drive motor
75 Port hydraulic reservoir
76 Centre fuselage integral fuel tank
77 Leading-edge flap drive hydraulic motor
78 Ammunition drum – 511 rounds
79 Upper position light/refuelling floodlight
80 TACAN aerial
81 Hydraulic accumulator
82 Starboard hydraulic reservoir
83 Leading-edge flap driveshaft
84 Inboard, No6 stores station (4,500lb/2,041kg capacity)
85 Pylon attachment hardpoint
86 Leading-edge flap driveshaft and rotary actuators
87 No7 stores hardpoint (700lb/318kg capacity)
88 Radar warning antenna
89 Missile launch rails
90 AMRAAM air-to-air missiles
91 Loading pod (carriage of essential ground equipment and personal effects for off-base deployment)
92 Starboard leading-edge manoeuvre flap, down position
93 Outboard No8 stores station hardpoint (700lb/318kg capacity)
94 Wing-tip No9 stores station (425lb/193kg capacity)
95 Wing-tip AMRAAM missile
96 Starboard navigation light
97 Fixed portion of trailing edge
98 Static dischargers
99 Starboard flaperon
100 Starboard wing integral fuel tank
101 Fuel system piping
102 Fuel pump
103 Starboard wing-root attachment attachment fishplates
104 Fuel tank access panels
105 Universal air refuelling receptacle (UARSSI), open
106 Engine compressor intake centrebody fairing
107 Airframe-mounted accessory-equipment gearbox
108 Jet fuel starter
109 Machined wing-attachment bulkheads
110 Engine fuel-management equipment
111 Pressure refuelling receptacle
112 Pratt & Whitney F100-PW-220 afterburning turbofan engine
113 VHF/IFF aerial
114 Starboard flaperon hydraulic actuator
115 Fuel-tank tail fins
116 Sidebody fairing integral fuel tank
117 Position light
118 Cooling-air ram air intake
119 Fin-root fairing
120 Forward engine support link
121 Rear fuselage integral fuel tank
122 Thermally insulated tank inner skin
123 Tank access panels
124 Radar-warning system power amplifier
125 Fin-root attachment fairings
126 Flight-control system hydraulic accumulators
127 Multi-spar fin construction
128 Starboard all-moving tailplane (tailplane panels interchangeable)
129 General Electric F110-GE-100 alternative power plant
130 Fin leading-edge honeycomb construction
131 Dynamic pressure sensor
132 Carbon-fibre fin skin panelling
133 VHF communications aerial (AM/FM)
134 Fin-tip antenna fairing
135 Anti-collision light
136 Threat-warning antenna
137 Static dischargers
138 Rudder honeycomb construction
139 Rudder hydraulic actuator
140 ECM antenna fairing
141 Tail navigation light
142 Extended tailcone fairing, brake-parachute housing (Norwegian aircraft) or Rapport III or Itek 69 ECM systems (Belgian or Israeli aircraft)
143 Variable-area afterburner nozzle flaps
144 Nozzle sealing fairing
145 Afterburner nozzle actuators (five)
146 Port split trailing-edge airbrake, open, upper and lower surfaces
147 Airbrake actuating linkage
148 Port all-moving tailplane
149 Static dischargers
150 Graphite-epoxy tailplane skin panels
151 Leading-edge honeycomb construction
152 Corrugated-aluminium sub-structure
153 Hinge pivot fixing
154 Tailplane hydraulic actuator
155 Fuel jettison chamber, port and starboard
156 Afterburner ducting
157 Rear fuselage machined bulkheads
158 Port position light
159 AN/ALE-40 (VO-4) chaff/flare dispenser, port and starboard
160 Main engine thrust mounting, port and starboard
161 Sidebody fairing frame construction
162 Runway arrester hook

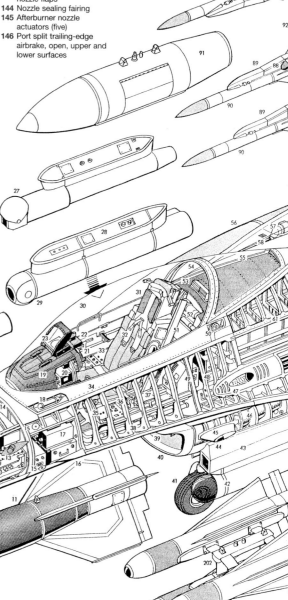

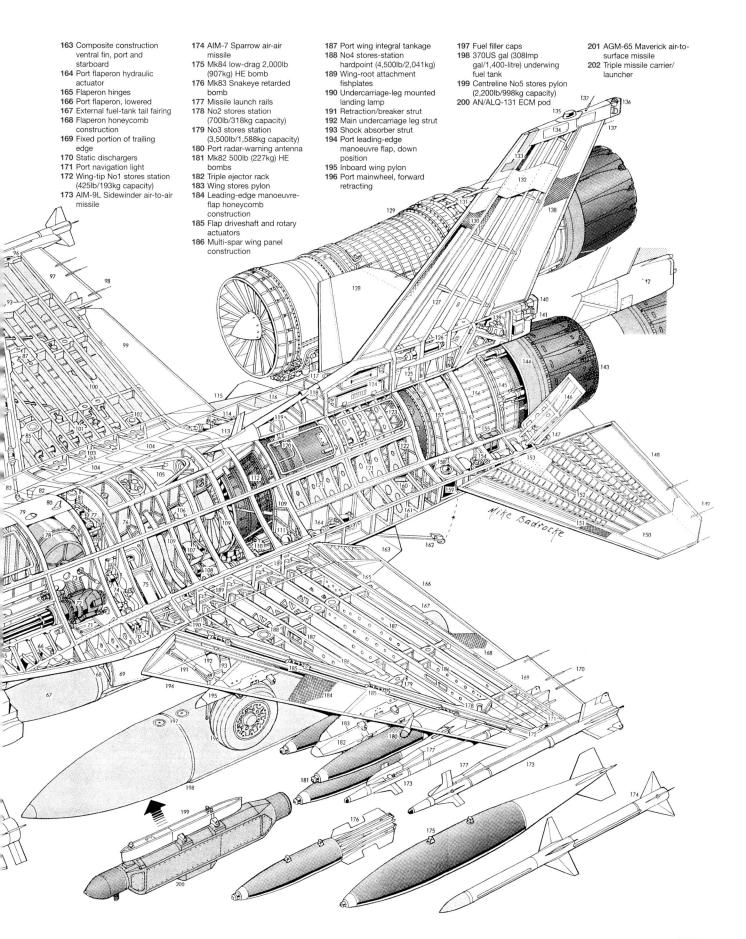

163 Composite construction ventral fin, port and starboard
164 Port flaperon hydraulic actuator
165 Flaperon hinges
166 Port flaperon, lowered
167 External fuel-tank tail fairing
168 Flaperon honeycomb construction
169 Fixed portion of trailing edge
170 Static dischargers
171 Port navigation light
172 Wing-tip No1 stores station (425lb/193kg capacity)
173 AIM-9L Sidewinder air-to-air missile

174 AIM-7 Sparrow air-air missile
175 Mk84 low-drag 2,000lb (907kg) HE bomb
176 Mk83 Snakeye retarded bomb
177 Missile launch rails
178 No2 stores station (700lb/318kg capacity)
179 No3 stores station (3,500lb/1,588kg capacity)
180 Port radar-warning antenna
181 Mk82 500lb (227kg) HE bombs
182 Triple ejector rack
183 Wing stores pylon
184 Leading-edge manoeuvre-flap honeycomb construction
185 Flap driveshaft and rotary actuators
186 Multi-spar wing panel construction

187 Port wing integral tankage
188 No4 stores-station hardpoint (4,500lb/2,041kg)
189 Wing-root attachment fishplates
190 Undercarriage-leg mounted landing lamp
191 Retraction/breaker strut
192 Main undercarriage leg strut
193 Shock absorber strut
194 Port leading-edge manoeuvre flap, down position
195 Inboard wing pylon
196 Port mainwheel, forward retracting

197 Fuel filler caps
198 370US gal (308Imp gal/1,400-litre) underwing fuel tank
199 Centreline No5 stores pylon (2,200lb/998kg capacity)
200 AN/ALQ-131 ECM pod

201 AGM-65 Maverick air-to-surface missile
202 Triple missile carrier/launcher

Mike Badrocke

Electrical System Controls and Indicators
C DF (Typical)

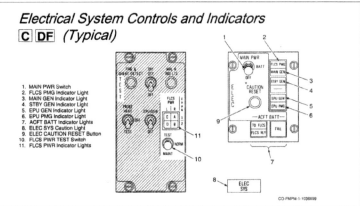

1. MAIN PWR Switch
2. FLCS PMG Indicator Light
3. MAIN GEN Indicator Light
4. STBY GEN Indicator Light
5. EPU GEN Indicator Light
6. EPU PMG Indicator Light
7. ACFT BATT Indicator Lights
8. ELEC SYS Caution Light
9. ELEC CAUTION RESET Button
10. FLCS PWR TEST Switch
11. FLCS PWR Indicator Lights

CO-FMPM-1-1038X99

CONTROL/INDICATOR	POSITION/INDICATION	FUNCTION
1. MAIN PWR Switch	MAIN PWR	Connects external power or the main generator to the electrical system and enables standby generator. Determines function of FLCS PWR TEST switch. If ac power is not available, connects aircraft battery to the battery buses
NOTE During ground operation, if the MAIN PWR switch is moved from MAIN PWR to OFF without a delay of 1 second in BATT, the EPU does not activate and electrical power for braking, NWS, hook, and radios is lost.	BATT	Connects aircraft battery to the battery buses, disconnects main generator or external power, resets main generator, disables standby generator, and determines function of FLCS PWR TEST switch
	OFF	In flight - disconnects main generator from electrical system and disables standby generator
		On ground - disconnects main generator or external power from aircraft electrical system and disables standby generator. Disconnects the aircraft battery from the battery buses. Canopy operation is available after engine shutdown
2. FLCS PMG Indicator Light	FLCS PMG (amber)	In flight - None of the FLCS branches are receiving power from the FLCS PMG
		On ground - FLCS PMG power is not available at one or more FLCS branches. Light is delayed 60 seconds after initial NLG WOW
3. MAIN GEN Indicator Light	MAIN GEN (amber)	Indicates external power or main generator not connected to one or both nonessential ac buses
4. STBY GEN Indicator Light	STBY GEN (amber)	Indicates standby generator power is not available
5. EPU GEN Indicator Light	EPU GEN (amber)	Indicates the EPU has been commanded on but the EPU generator is not providing power to both emergency ac buses. The light does not function with the EPU switch in OFF (WOW) and the engine running
6. EPU PMG Indicator Light	EPU PMG (amber)	Indicates the EPU has been commanded on but EPU PMG power is not available to all branches of the FLCS
7. ACFT BATT Indicator Lights	FAIL (amber)	In flight - indicates aircraft battery failure (20V or less)
		On ground - indicates aircraft battery or battery charger failure. Light is delayed 60 seconds after MLG WOW
	TO FLCS (amber)	In flight - indicates battery bus power is going to one or more FLCS branches and voltage is 25V or less
		On ground - indicates battery bus power is going to one or more FLCS branches
	FLCS RLY (amber)	Indicates that voltage on one or more of the four FLCC branches connected to the aircraft battery is inadequate (below 20V) or that one or more FLCC branches are not connected to the battery
8. ELEC SYS Caution Light	ELEC SYS (amber)	Illuminates in conjunction with any of the above lights
9. ELEC CAUTION RESET Button	Push	Resets resettable overcurrent protection units and ELEC SYS caution light and clears MASTER CAUTION light for future indications. Resets main and standby generators
10. FLCS PWR TEST Switch	TEST	When MAIN PWR switch is in:
		MAIN PWR / **BATT**
		Tests FLCC power output / Tests FLCC power output on aircraft battery
	NORM	Normal position. Tests EPU PMG power availability during EPU/GEN test on ground / NA
	MAINT	For maintenance use on the ground. Inoperative in flight
11. FLCS PWR Indicator Lights	A, B, C, and D (green)	Illuminate to indicate proper power output of FLCC during FLCS power tests

ABOVE The F-16CM pilot interfaces the electrical systems via a simple set of controls and indicators. However, knowledge of the entire system must be flawless, not least because the loss of a bus or circuit can cause key systems to cease functioning or to function in a degraded state. *(USAF)*

The aircraft battery can provide temporary emergency power to the FLCS, although the charge level of the battery determines for how long.

The FLCS incorporates four latching relays that prevent the FLCC from being connected to the aircraft battery until the aircraft's jet fuel starter (JFS – see Chapter 6) is initiated. The relays therefore prevent the battery from being drained during maintenance operations. For maintenance purposes, external power provisions include a standard external power cable receptacle and a monitor unit. The latter allows external power to be connected to the aircraft buses, providing the same power as the main generator would if it were online.

Hydraulic system

Hydraulic pressure is supplied by two 3,000psi hydraulic systems designated systems A and B. These are powered by two independent engine-driven pumps located on the ADG. Each system has a pressurised reservoir to store hydraulic fluid.

Systems A and B operate simultaneously to supply hydraulic power for the primary flight controls and leading-edge flaps. If one system fails, the remaining system provides sufficient hydraulic pressure to actuate the flight controls, albeit at a reduced rate.

System A also supplies power to the fuel flow proportioner and the speedbrakes. Meanwhile, system B powers all remaining utility functions, including the M61A1 and gun purge door, air refuelling system, landing gear, brakes, nose-wheel steering and JFS accumulators (which provide start power for the JFS and backup pressure for the brakes).

The landing gear can be extended pneumatically in the event of hydraulic system B failure. If both systems fail, a third hydraulic pump located on the EPU automatically provides hydraulic pressure to system A.

Each hydraulic system has FLCS accumulator that, when demand exceeds the pump maximum flow rate during rapid control surface movement, provides additional hydraulic flow. In the event of both hydraulic systems failing, the two FLCS accumulators provide adequate hydraulic pressure to the

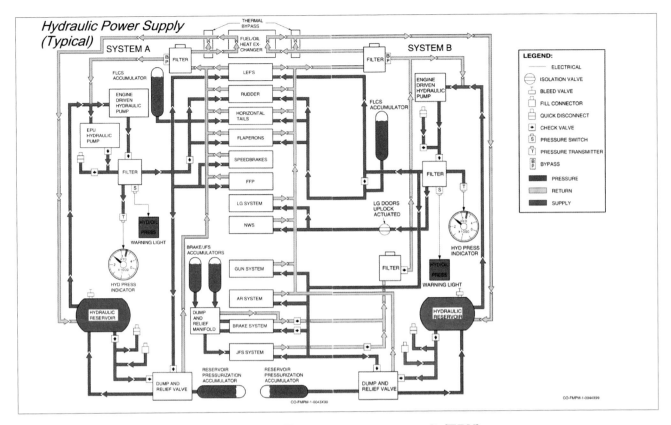

Hydraulic Power Supply (Typical)

flight controls until the EPU reaches the required speed.

HYD PRESS (pressure) indicators, one for each system, are located on the right auxiliary console and are powered by emergency ac bus No2. A HYD/OIL PRESS warning light comes on when hydraulic pressure in either system drops below 1,000psi or when engine oil pressure drops below 10(±2)psi. The light is powered by battery bus No1.

Emergency power unit (EPU)

The EPU is a self-contained unit that simultaneously provides emergency hydraulic pressure to system A and emergency electrical power. It uses engine bleed air and/or hydrazine to operate, and automatically activates when both main and standby generators fail, or when both hydraulic system pressures fall below 1,000psi. However, it can be manually engaged by the pilot at any time.

ABOVE The hydraulic system runs at 3,000psi and is split into systems A and B. Where possible redundancy has been built in, as this schematic shows. *(USAF)*

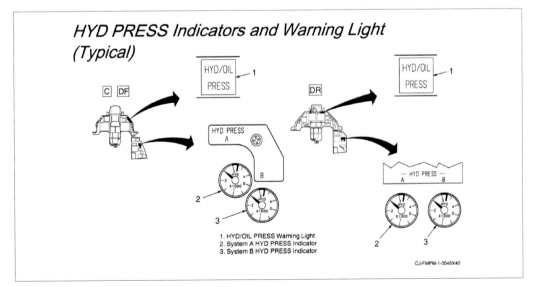

HYD PRESS Indicators and Warning Light (Typical)

1. HYD/OIL PRESS Warning Light
2. System A HYD PRESS Indicator
3. System B HYD PRESS Indicator

LEFT The pilot's hydraulic indicators consist of warning lights and pressure dials – the latter often being referred to as 'peanut' gauges as a result of their small size. *(USAF)*

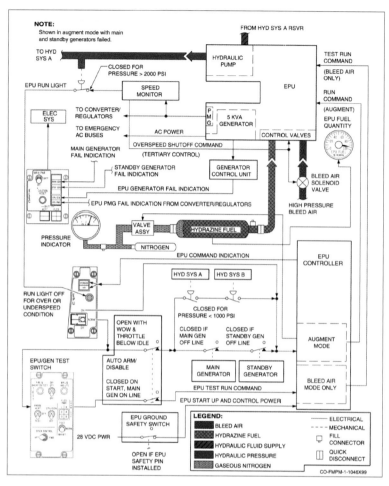

The EPU flow diagram. Labels include:

NOTE: Shown in augment mode with main and standby generators failed.

FROM HYD SYS A RSVR · HYDRAULIC PUMP · TEST RUN COMMAND · (BLEED AIR ONLY) · RUN COMMAND · (AUGMENT) · EPU FUEL QUANTITY · TO HYD SYS A · CLOSED FOR PRESSURE > 2000 PSI · EPU RUN LIGHT · SPEED MONITOR · EPU · ELEC SYS · TO CONVERTER/ REGULATORS · TO EMERGENCY AC BUSES · AC POWER · PMG · 5 KVA GENERATOR · CONTROL VALVES · OVERSPEED SHUTOFF COMMAND · (TERTIARY CONTROL) · MAIN GENERATOR FAIL INDICATION · STANDBY GENERATOR FAIL INDICATION · GENERATOR CONTROL UNIT · BLEED AIR SOLENOID VALVE · EPU GENERATOR FAIL INDICATION · HIGH PRESSURE BLEED AIR · EPU PMG FAIL INDICATION FROM CONVERTER/REGULATORS · VALVE ASSY · HYDRAZINE FUEL · PRESSURE INDICATOR · NITROGEN · EPU COMMAND INDICATION · EPU CONTROLLER · RUN LIGHT OFF FOR OVER OR UNDERSPEED CONDITION · HYD SYS A · HYD SYS B · CLOSED FOR PRESSURE < 1000 PSI · OPEN WITH WOW & THROTTLE BELOW IDLE · CLOSED IF MAIN GEN OFF LINE · CLOSED IF STANDBY GEN OFF LINE · AUGMENT MODE · EPU/GEN TEST SWITCH · AUTO ARM/ DISABLE · MAIN GENERATOR · STANDBY GENERATOR · BLEED AIR MODE ONLY · CLOSED ON START, MAIN GEN ON LINE · EPU TEST RUN COMMAND · EPU START UP AND CONTROL POWER · EPU GROUND SAFETY SWITCH · 28 VDC PWR · OPEN IF EPU SAFETY PIN INSTALLED

LEGEND:
- BLEED AIR
- HYDRAZINE FUEL
- HYDRAULIC FLUID SUPPLY
- HYDRAULIC PRESSURE
- GASEOUS NITROGEN
- ELECTRICAL
- MECHANICAL
- FILL CONNECTOR
- QUICK DISCONNECT

CO-FMPM-1-1046X99

ABOVE The EPU flow diagram shows the hydrazine fuel reservoir in red in the centre. The EPU has saved many an F-16 pilot from having to eject. (USAF)

The EPU requires dc power from either of the two battery buses to start, and will in turn provide power to the emergency ac and dc buses once it is running. Nonessential and essential dc buses are left unpowered whenever

the EPU is operating, thereby reducing electrical loads. If the normal system A hydraulic pump fails, the EPU is the only source of system A pressure.

Hydrazine is used whenever the EPU is commanded to start, except when activated during ground operation. When the EPU starts, hydrazine is forced by nitrogen pressure into a decomposition chamber. The resultant gases spin the turbine/gearbox, which then powers the EPU generator and hydraulic pump.

Hydrazine exhaust is vented overboard (lower inboard side of the right ventral strake) and consists primarily of nitrogen, hydrogen, ammonia and water. The temperature of exhaust gases can reach 1,600°F and is highly flammable. Once started, engine bleed air is normally used to maintain operating APU speed. However, when bleed air is insufficient hydrazine is used.

Landing gear (LG) system

The nose landing gear (NLG) is extended and retracted by hydraulic pressure from system B, while the main landing gears (MLGs) are retracted hydraulically but extended by gravity and air loads.

All LG doors are hydraulic, employing electrical sequencing during retraction and mechanical sequencing during extension. If hydraulic system B fails the LG may be extended pneumatically. The MLG tyres are certified for use to 225kt ground speed; the NLG tyre is certified for use to 217kt ground speed.

The two MLGs are independent of each other. They combine a forward retraction with a mechanical wheel twist, ending in the gear fitting into two separate wheel wells. Each MLG wheel is equipped with three thermal pressure relief plugs that allow the wheel to deflate slowly if the tyre overheats.

The NLG retracts aft, mechanically twisting through 90° as it does so. A torque arm quick-disconnect allows the nose wheel to be turned beyond the steerable range for towing on the ground.

The LG control panel is located on the cockpit's left auxiliary console. It consists of a wheel-shaped landing gear handle that operates electrical switches to command LG retraction

RIGHT In the F-16CM cockpit, the pilot controls the EPU through a simple panel. The guarded switch can be used to override the automatic functioning of the unit. (USAF)

1. HYDRAZN Light (Amber)
2. AIR Light (Amber)
3. EPU Run Light (Green)
4. EPU Switch

CO-FMPM-1-0047X99

(up position) or extension (down position).

Three green 'WHEELS down' lights, located on the LG control panel, illuminate when the respective landing gear is down and locked, and extinguish when the gear is up and safely locked into place.

A warning light in the handle illuminates when the LG and associated doors are in transit or have failed to lock in the commanded position. The warning light also illuminates when the following conditions are met: one or more LG are not down and locked, airspeed is less than 190kt, altitude is less than 10,000ft, and rate of descent is greater than 250ft per minute.

To prevent retraction of the landing gear the handle is locked in the down position by a 'weight on wheels' (WoW) switch when the aircraft is on the ground. Once airborne, this 'WoW' switch activates a solenoid that unlocks the handle. Similarly, once in the UP position, the handle is secured by a spring-activated

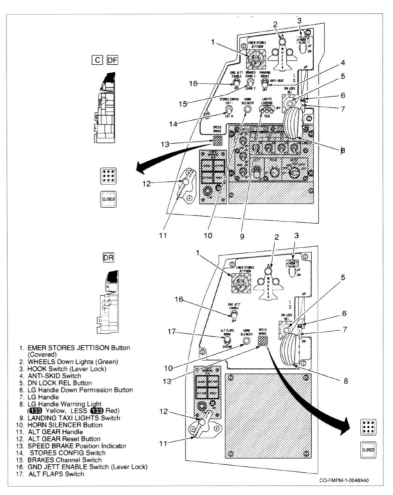

1. EMER STORES JETTISON Button (Covered)
2. WHEELS Down Lights (Green)
3. HOOK Switch (Lever Lock)
4. ANTI-SKID Switch
5. DN LOCK REL Button
6. LG Handle Down Permission Button
7. LG Handle
8. LG Handle Warning Light (**133** Yellow, LESS **133** Red)
9. LANDING TAXI LIGHTS Switch
10. HORN SILENCER Button
11. ALT GEAR Handle
12. ALT GEAR Reset Button
13. SPEED BRAKE Position Indicator
14. STORES CONFIG Switch
15. BRAKES Channel Switch
16. GND JETT ENABLE Switch (Lever Lock)
17. ALT FLAPS Switch

CO-FMPM-1-0048X40

LEFT A maintainer services an F-16 main landing gear wheel. The Viper's narrow-track undercarriage can make it unforgiving during landing. *(USAF)*

BELOW Cleanliness may well be next to Godliness, but for the F-16 crew chief, a clean undercarriage is the best way of ensuring that any hydraulic leaks are spotted quickly. *(USAF)*

lock to prevent LG extension during high g manoeuvres. To release the solenoid and allow the handle to be positioned into the 'DN' (down) position, the LG 'handle down' permission button (located on the gear handle stalk) must first be depressed.

As a failsafe, a 'DN LOCK REL' (down lock release) button mechanically unlocks the spring-actuated handle lock if the landing gear down position handle becomes inoperative or unpowered. The 'DN LOCK REL' will also override the 'WoW' switch, meaning that the gear handle can be moved to the UP position whilst on the ground – this will result in landing gear retraction.

If normal extension of the gear is for some reason impossible, an 'ALT GEAR' handle can be pulled to supply pneumatic pressure to all LG doors. This will result in the extension of the NLG and MLG.

To prevent inadvertent wheels-up landings, and to alert the pilot to landing gear anomalies, a warning horn will sound when the NLG or MLG is not down and locked and all the following three conditions exist: airspeed is below 190kt, pressure altitude is less than 10,000ft, and the rate of descent is greater than 250fpm. The pilot silences the horn by depressing the 'HORN SILENCER' button.

When the same conditions exist that trigger the warning horn, the 'TO/LDG CONFIG' (take-off/landing configuration) warning light (located on the right glare-shield) will also illuminate if the trailing-edge flaps are not fully down and/or the NLG or either MLG are not down and locked.

In addition, the 'TO/LDG CONFIG' warning light illuminates on the ground when the trailing-edge flaps are not fully down.

Lowering and raising the gear handle automatically lowers or retracts (respectively) the trailing-edge flaps. It also commands the FLCS to enter the appropriate mode for that phase of flight – raising the handle results in the FLCS's cruise mode being entered, for example.

A nose-wheel steering (NWS) system, electrically controlled using dc bus No1, is hydraulically operated using system B pressure. Controlled by the rudder pedals, the NWS is limited to 32° in each direction. NWS is automatically disengaged when the NLG strut is fully extended, but is otherwise controlled by the 'NWS AIR DISC MSL STEP' (NWS/air refuelling disconnect) button on the side of the control stick.

An 'AR/NWS' status indicator light, located on the top of the glare shield, illuminates green when NWS is engaged. For failures, an 'NWS FAIL' light illuminates on the caution light panel.

Wheel brake system

Each of the F-16's MLG wheels is equipped with a hydraulically powered multiple disc brake. The brakes are electrically controlled by toe brake pedals, and the amount of braking gradually increases as pedal pressure is applied. The brakes may be applied singly or simultaneously from the forward or rear cockpits.

Brake hydraulic power is supplied by system B. If system B fails or the engine is operating at less than 12% rpm, the toe brakes and parking brake are available until the brake/JFS accumulators deplete. A parking brake is provided, as is an anti-skid system that protects against blown tyres.

The parking brake is activated by the 'ANTI-SKID' switch located on the LG control panel, and supplies full, unmetered pressure to three of the six pistons in each brake. It is powered by battery bus No2 and system B hydraulics or one brake/JFS accumulator.

The toe brakes use electrical power from the FLCC and CHAN I and CHAN 2 dc power sources. The pedal signals are supplied to the brake control/anti-skid assembly, which relies

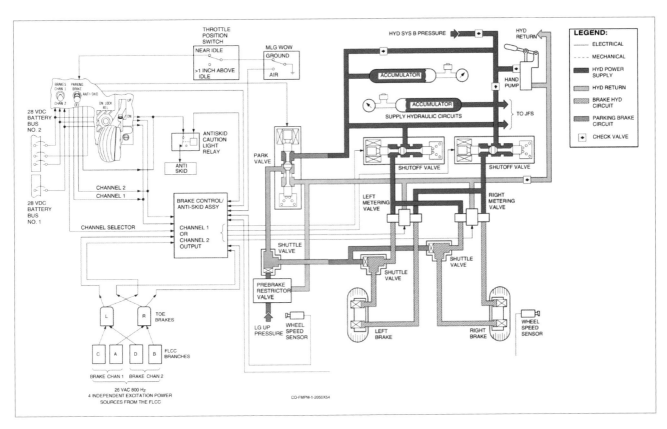

on CHAN 1 and CHAN 2 dc power sources to operate valves for controlling hydraulic pressure to the brakes. CHAN 1 and CHAN 2 are powered by battery buses No1 and No2 respectively.

Hydraulic pressure to three of the six pistons in each brake is controlled by electrical power from one dc power source and pressure to the other three pistons of each brake is controlled by electrical power from another dc power source.

Anti-skid is available in either brake channel any time the toe brakes are powered. It will deliver a corresponding deceleration rate to a given pedal deflection. The deceleration skid control will dampen brake pedal inputs to the brakes resulting in a smooth, efficient stop.

The anti-skid system incorporates touchdown skid control that prevents brake application prior to wheel spin-up, even if brake pedals are fully depressed. It also offers deceleration skid control, which is active when either brake pedal deflection is less than 85% of maximum and the runway surface can provide the requested deceleration (ie is not wet or slippery). Maximum performance skid control becomes active when both brake pedal deflections are equal to or greater than 85% or deceleration skid control is ineffective because the runway surface cannot provide requested deceleration. An anti-skid failure detection will illuminate the 'ANTI-SKID' caution light.

ABOVE The Viper has a sophisticated wheel-brake system powered in normal operation by hydraulic system B. The system responds to the amount of toe-brake travel applied by the pilot. *(USAF)*

LEFT The multiple disc brake on the F-16 is hydraulically powered and electronically controlled in response to the pilot's toe-brake inputs. *(USAF)*

An alternate braking mode cuts in when failure of a wheel-speed sensor is detected. In this mode, if differential braking is applied (15% or greater difference between pedals), both brakes alternate between pedal pressure as metered and no pressure. Braking effectiveness is reduced by 50% or greater. If brake pedals are within 15%, the system uses the information from the remaining wheel speed sensor and stopping distance is increased by approximately 25% on both wet and dry surfaces.

The 'ANTI-SKID' switch, located on the LG control panel, controls the following: 'PARKING BRAKE' – full unmetered brake pressure is applied with the throttle in the 'OFF' to 'IDLE' range (advancing the throttle 1in beyond 'IDLE' automatically returns the switch to 'ANTI-SKID', which in turn releases the parking brake); 'ANTI-SKID' – anti-skid protection is energised; and 'OFF'.

Speedbrake system

The speedbrake system consists of two pairs of clamshell doors located on each side of the engine nozzle and inboard of the horizontal tail. They are powered by hydraulic system A.

The speedbrakes open to 60° with the right MLG not down and locked. With the right MLG down and locked, speedbrake opening is limited to 43°, thereby helping to prevent the bottom of the clamshell surfaces from grinding the runway during landing. Holding the 'SPD BRK' switch on the throttle in the open (aft) position will override this. When the NLG strut compresses on landing, the speedbrakes can be fully opened.

The 'SPD BRK' switch is located on the throttle and is a three-position slide switch. The open (aft) position is spring-loaded to off (centre) and allows the speedbrakes to be incrementally opened. The closed (forward) position has a detent that holds the switch in place, allowing a single motion to close the speedbrakes. In the two-seat F-16D, the speedbrake switches are connected in parallel and function so that either can override the other by holding in the open position. If one switch is in the closed position, the speedbrakes close when the other is released from the open position.

A three-position 'SPEED BRAKE' indicator is located on the LG control panel. It shows the following indications: 'CLOSED' – both speedbrakes closed; speedbrake symbol – speedbrakes are not closed; and diagonal lines – electrical power removed from the indicator, but they also appear momentarily when opening or closing the speedbrakes.

Arrestment system

The F-16 features a simple arrestment system for snagging a runway arrestment cable in emergencies. Consisting of an electrically

RIGHT The clamshell doors of the F-16's speedbrakes – seen here in their fully open position – are controlled by a thumb-switch on the throttle. It is not unheard of for a pilot to drag them along the ground during aero braking on landing! *(USAF)*

controlled hook, it is operated pneumatically. Pressure for extension is supplied by the LG/ hook emergency pneumatic bottle. Once extended the hook cannot be fully retracted without ground crew assistance, but can be raised enough to disengage the cable.

The 'HOOK' switch, located on the LG control panel, is lever-locked in the 'UP' or 'DN' position. Positioning the switch to 'DN' causes the hook to extend. Returning the switch to 'UP' partially retracts the hook, allowing for cable disengagement and for taxi over the cable. A 'HOOK' caution light, located on the caution light panel, illuminates any time the hook is not up and locked.

Wing flap system

The LEFs (leading-edge flaps) consist of a spanwise flap on each wing leading edge controlled as a function of Mach number, angle of attack and altitude, by command signals from the FLCC.

An asymmetry sensing and braking mechanism prevents LEF asymmetry. If an asymmetry is sensed, the LEF's 'FLCS LEF LOCK PFL' caution is displayed, and the 'FLCS' warning light illuminates. The LEFs are automatically programmed when the

'LE FLAPS' switch is in the 'AUTO' position. However, when weight is on both MLGs, the LEFs are positioned 2° up, as they are when the throttle is at 'IDLE' and MLG wheel speed is greater than 60kt ground speed.

The flaperons (trailing edge flaps) are located on the wing trailing edge and function as both ailerons and as TEFs. The flaperons have a maximum command deflection of 20° down and 23° up . When acting as flaps the deflection is downward, but when acting as ailerons the deflection can be either up or down. Both functions are operable whenever the FLCS is powered.

The TEFs are controlled as a function of the LG handle position, the 'ALT FLAPS' switch, airspeed and Mach number. Positioning the LG handle to 'DN' or the 'ALT FLAPS' switch to 'EXTEND' causes the TEFs to deflect downward. At all airspeeds below 240kt the TEF position is 20° down. Above 240kt the TEFs reduce deflection as a function of airspeed until nearly fully retracted at 370kt.

The 'ALT FLAPS' switch features a 'NORM' position (the TEFs are controlled by the LG handle and airspeed), and an 'EXTEND' position (lowers the TEFs, depending on airspeed). The switch does not affect the operation of the LEFs unless the FLCS is operating on standby gains.

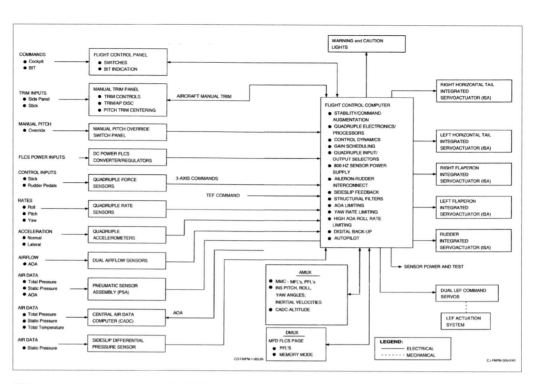

Flight control system (FLCS)

The FLCS is a digital four-channel, fly-by-wire system that hydraulically positions the flight control surfaces. Electrical signals are generated through the stick and rudder pedals.

A main component of the FLCS is the flight control computer (FLCC). Redundancy is provided in electronic branches, hydraulic systems, power supplies and sensor systems.

Command signals to the FLCC are initiated by applying force to the stick and rudder pedals. These signals are processed by the FLCC, along with signals from the air data

system, flight control rate gyros, accelerometers and INS. The processed signals are transmitted and used to command the horizontal tails, flaperons and rudder.

Pitch is controlled by symmetrical movement of the horizontal tails. Roll is controlled by differential movement of the flaperons and horizontal tails. Yaw is controlled by the rudder. Roll coordination is provided by an ARI (aileron rudder interconnect). The ARI is not available whenever MLG wheel speed exceeds 60kt or if angle of attack exceeds 35°.

After take-off, ARI is activated within two seconds of the LG handle being raised (the delay comes from waiting for the spin-down braking system to stop the main wheels). If the LG handle remains down, 10–20 seconds are required for the MLG wheels to spin down and activate ARI.

A digital backup (DBU) provides a software backup in the event of software problems in the primary program. The DBU is a reduced set of control laws that automatically engage when software problems in the FLCC force a majority of the branches into a failed state. DBU can only be disengaged by use of the 'DIGITAL BACKUP' switch.

FLCS limiters are provided in all three axes to help prevent departures from controlled flight. These are: AoA/G limiter, roll-rate limiter, rudder authority limiter and yaw rate limiter.

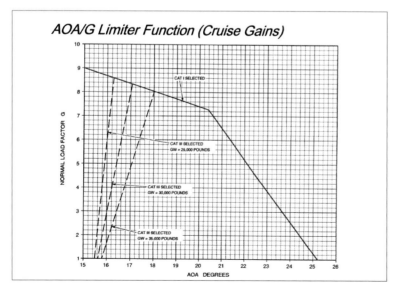

The AoA/G limiter reduces the positive g available as a function of AoA when the FLCS is operating in cruise gains (see 'FLCS gains' below). The negative g available is a function of airspeed. Below 15° AoA, the maximum positive g available is +9g. As AoA increases, the maximum allowable positive g decreases. The positive g limit and maximum AoA depend on the position of the 'STORES CONFIG' switch.

With the 'STORES CONFIG' switch in the CAT I position, positive g decreases to a value of 1g at 25° AoA. Maximum commanded AoA is approximately 25.5°. In CAT III, maximum AoA varies from approximately 16–18° as a function of gross weight and g. The negative g available above approximately 250kt is -3g. Below 250kt the available negative g varies between -3g and zero g as a function of airspeed, altitude and AoA.

In take-off and landing gains, the 'STORES CONFIG' switch has no effect on limiting or gains. Maximum positive g is a function of airspeed and AoA, and the negative g available is fixed. The maximum AoA for 1g is approximately 21°. In inverted or upright departures, the AoA/g limiter will override stick pitch commands if the MPO is not engaged. The MPO can always override the negative g function of the limiter. It can also override the AoA function of the limiter when the AoA exceeds 35°.

In cruise gains, a roll-rate limiter reduces available roll-rate authority to help prevent roll-coupled departures. This authority is reduced as airspeed decreases, AoA increases, or trailing edge down horizontal tail deflection increases. Roll authority is further reduced for large total rudder commands.

In take-off and landing gains, roll-rate limiting is available but is a fixed value independent of AoA, airspeed, or horizontal tail position.

The rudder authority limiter reduces the pedal-commanded rudder deflection as a function of AoA, roll rate, and 'STORES CONFIG' switch position in cruise gains. However, ARI authority, stability augmentation and trim authority are not reduced. In take-off and landing gains category I rudder authority limiting is provided.

When AoA exceeds 35° the yaw-rate limiter overrides pilot roll and rudder commands and

provides flaperons with and rudder against the yaw rate until AoA is below 32° to enhance spin resistance. The yaw rate limiter provides no protection against yaw departures in the normal flying range (-5–25° AoA). When AoA decreases below -5° and airspeed is less than 170kt the yaw rate limiter engages but does not affect pilot roll and rudder commands. Pilot roll and rudder commands are inhibited during inverted departures only when the MPO is engaged. The yaw-rate limiter provides rudder against the yaw rate until AoA is above -5° to enhance spin resistance.

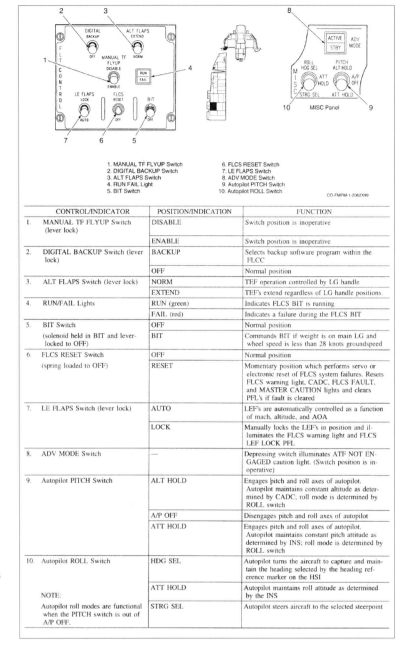

1. MANUAL TF FLYUP Switch
2. DIGITAL BACKUP Switch
3. ALT FLAPS Switch
4. RUN FAIL Light
5. BIT Switch
6. FLCS RESET Switch
7. LE FLAPS Switch
8. ADV MODE Switch
9. Autopilot PITCH Switch
10. Autopilot ROLL Switch

CO-FMPM-1-2062X99

	CONTROL/INDICATOR	POSITION/INDICATION	FUNCTION
1.	MANUAL TF FLYUP Switch (lever lock)	DISABLE	Switch position is inoperative
		ENABLE	Switch position is inoperative
2.	DIGITAL BACKUP Switch (lever lock)	BACKUP	Selects backup software program within the FLCC
		OFF	Normal position
3.	ALT FLAPS Switch (lever lock)	NORM	TEF operation controlled by LG handle
		EXTEND	TEF's extend regardless of LG handle positions.
4.	RUN/FAIL Lights	RUN (green)	Indicates FLCS BIT is running
		FAIL (red)	Indicates a failure during the FLCS BIT
5.	BIT Switch (solenoid held in BIT and lever-locked to OFF)	OFF	Normal position
		BIT	Commands BIT if weight is on main LG and wheel speed is less than 28 knots groundspeed
6.	FLCS RESET Switch (spring loaded to OFF)	OFF	Normal position
		RESET	Momentary position which performs servo or electronic reset of FLCS system failures. Resets FLCS warning light, CADC, FLCS FAULT, and MASTER CAUTION lights and clears PFL's if fault is cleared
7.	LE FLAPS Switch (lever lock)	AUTO	LEF's are automatically controlled as a function of mach, altitude, and AOA
		LOCK	Manually locks the LEF's in position and illuminates the FLCS warning light and FLCS LEF LOCK PFL
8.	ADV MODE Switch	—	Depressing switch illuminates ATF NOT ENGAGED caution light. (Switch position is inoperative)
9.	Autopilot PITCH Switch	ALT HOLD	Engages pitch and roll axes of autopilot. Autopilot maintains constant altitude as determined by CADC; roll mode is determined by ROLL switch
		A/P OFF	Disengages pitch and roll axes of autopilot
		ATT HOLD	Engages pitch and roll axes of autopilot. Autopilot maintains constant pitch attitude as determined by INS; roll mode is determined by ROLL switch
10.	Autopilot ROLL Switch	HDG SEL	Autopilot turns the aircraft to capture and maintain the heading selected by the heading reference marker on the HSI
	NOTE:	ATT HOLD	Autopilot maintains roll attitude as determined by the INS
	Autopilot roll modes are functional when the PITCH switch is out of A/P OFF.	STRG SEL	Autopilot steers aircraft to the selected steerpoint

ABOVE The Flight Control panel provides the interface between pilot and FLCS. Direct controls are provided over leading-edge flaps and automatic pitch up ('FLYUP') during terrain-following flight, in addition to the autopilot modes. *(USAF)*

FLCS gains

During normal operation, the FLCS receives inputs (gains) from the Air Data Converter (ADC) and provides relatively constant aircraft response for a given stick input, regardless of altitude or airspeed. This response varies slightly depending on configuration. In the event of a dual air data failure, the FLCS switches to standby (fixed) gains.

The FLCS is in cruise gains with the LG handle in 'UP', the 'ALT FLAPS' switch either in 'NORM' or in 'EXTEND' above 400kt, and the 'AIR REFUEL' switch either in 'CLOSE' or in 'OPEN' above 400kt. At low AoA the pitch axis of the FLCS is a g command system. As AoA increases the FLCS switches to a blended g and AoA system to provide a warning of high AoA and/or low airspeed. Roll rate limiting is available and maximum roll rate decreases as a function of low airspeed, high AoA and horizontal tail position.

The FLCS is in take-off and landing gains with the LG handle in 'DN', the 'ALT FLAPS' switch in 'EXTEND' (below 400kt), or the 'AIR REFUEL' switch in 'OPEN' (below 400kt). In take-off and landing gains the FLCS pitch axis operates as a pitch rate command system until 10° AoA and a blended pitch rate and AoA command system above 10° AoA. Roll rate limiting is available but is a fixed value independent of AoA, airspeed or horizontal tail position.

In standby gains, control response is tailored for a fixed altitude (sea level, standard day) and airspeed (LG handle in 'UP', approximately 600kt; LG handle in 'DN', approximately 230kt). The 'FLCS' warning light and 'FLCS FAULT' caution light illuminate. When operating on standby gains the LEFs are at 0° with the LG handle in UP and the 'ALT FLAPS' switch in 'NORM'. The LEFs deflect 15° down with the LG handle in 'DN' or the 'ALT FLAPS' switch in 'EXTEND'.

An FLCS data recorder is attached to the ejection seat. It retains the same information as the FLCC, including FLCS failure data, airspeed, altitude, true heading and elapsed time from take-off.

An angle-of-sideslip feedback function provides improved departure prevention by using AoS (angle of sideslip) and AoS rate feedback to position the control surfaces (primarily the rudder) to reduce sideslip.

The FLCS also automatically compensates for the off-centre gun and the aerodynamic effects of gun gas emissions during firing. It does so through small flaperon and rudder movements. Gun compensation is optimised for 0.7–0.9 Mach range.

FLCS controls

The side stick controller, mounted on the pilot's right, is attached to a force-sensing unit consisting of transducers that measure input in pitch and roll axes. The stick moves approximately ¼in in both axes, and is rotated slightly clockwise. Maximum nose-up and nose-down pitch commands are generated by 25lb and 16lb of input respectively. Maximum roll commands are generated by 17lb of side pressure in cruise gains, and by 12lb in take-off and landing gains.

Wrist rest and armrest assemblies are

BELOW A flow diagram showing the inputs taken for all three axes by the FLCC before being converted into flight-control surface movements. (USAF)

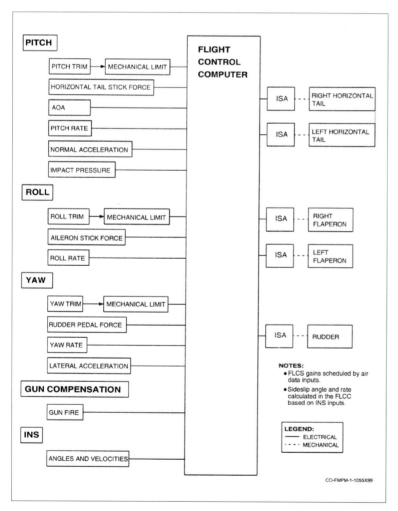

PITCH
- PITCH TRIM → MECHANICAL LIMIT
- HORIZONTAL TAIL STICK FORCE
- AOA
- PITCH RATE
- NORMAL ACCELERATION
- IMPACT PRESSURE

ROLL
- ROLL TRIM → MECHANICAL LIMIT
- AILERON STICK FORCE
- ROLL RATE

YAW
- YAW TRIM → MECHANICAL LIMIT
- RUDDER PEDAL FORCE
- YAW RATE
- LATERAL ACCELERATION

GUN COMPENSATION
- GUN FIRE

INS
- ANGLES AND VELOCITIES

FLIGHT CONTROL COMPUTER

- ISA — RIGHT HORIZONTAL TAIL
- ISA — LEFT HORIZONTAL TAIL
- ISA — RIGHT FLAPERON
- ISA — LEFT FLAPERON
- ISA — RUDDER

NOTES:
- FLCS gains scheduled by air data inputs.
- Sideslip angle and rate calculated in the FLCC based on INS inputs.

LEGEND:
— ELECTRICAL
- - - - MECHANICAL

CO-FMPM-1-1055X99

RIGHT The manual trim panel features rotary dials that allow the pilot to override FLCS trimming. This can sometimes be necessary when carrying an asymmetric stores load. *(USAF)*

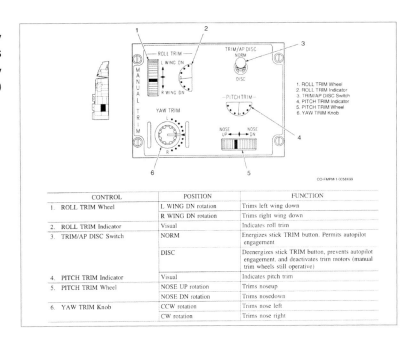

	CONTROL	POSITION	FUNCTION
1.	ROLL TRIM Wheel	L WING DN rotation	Trims left wing down
		R WING DN rotation	Trims right wing down
2.	ROLL TRIM Indicator	Visual	Indicates roll trim
3.	TRIM/AP DISC Switch	NORM	Energizes stick TRIM button. Permits autopilot engagement
		DISC	Deenergizes stick TRIM button, prevents autopilot engagement, and deactivates trim motors (manual trim wheels still operative)
4.	PITCH TRIM Indicator	Visual	Indicates pitch trim
5.	PITCH TRIM Wheel	NOSE UP rotation	Trims noseup
		NOSE DN rotation	Trims nosedown
6.	YAW TRIM Knob	CCW rotation	Trims nose left
		CW rotation	Trims nose right

provided in the form of adjustable brackets on the right sidewall, aft of the stick.

The rudder pedals are attached to force-sensing units containing transducers, each of which produces electrical yaw command signals. They also generate brake and NWS signals. Pedal feel is provided by mechanical springs.

Although the FLCS automatically trims the F-16 for 1g flight, controls on the 'MANUAL TRIM' panel (located on the left console) allow the pilot to trim pitch, roll and yaw manually. Typically it is used when the aircraft has an asymmetric stores configuration, or when the rudder is mistrimmed by the FLCS.

A 'MANUAL PITCH' override (MPO) switch, located on the left console, has two positions, 'NORM' and 'OVRD', and is spring-loaded to the 'NORM' position. If the aircraft enters a deep stall condition, the switch is positioned and held in 'OVRD', at which point the negative g limiter is disengaged. If AoA exceeds 35° the 'OVRD' position overrides the AoA/g limiter and allows pitch commands.

The 'STORES CONFIG' switch, located on the LG control panel, is set by the pilot according to the weight and g limits of external stores and the aircraft pylons. It has two positions: CAT I and CAT III. The CAT III position is selected when the aircraft is configured with a category III loading. CAT III provides additional AoA and g limiting to prevent stores and pylons from being overstressed, and to mitigate against departure from controlled flight.

	PITCH AXIS	ROLL AXIS	YAW AXIS
CAT I	Maximum AOA=25° g command system until 15° AOA g/AOA command system above 15° AOA	Maximum roll rate command decreases with: • AOA above 15° • Airspeed less than 250 knots • Horizontal tail deflection more than 5° trailing edge down • Total rudder command (from pilot and FLCS) exceeding 20° • Combination of horizontal tail greater than 15° trailing edge down, and AOA above 22°	Maximum deflection (pedal command) reduced for: • AOA>14° (zero roll rate) • Roll rate>20°/sec NOTE: Zero rudder authority available at 26° AOA.
CAT III	Maximum AOA = 16° - 18° (depending on GW) g command system until 7° AOA at 100 knots to 15° AOA at 420 knots and above g/AOA command system above these values	Maximum roll rate command reduced by approximately 40 percent of CAT I authority. Additional decreases as function of AOA, airspeed, horizontal tail position, and total rudder command	Maximum deflection (pedal command) reduced for: • AOA>3° (zero roll rate) • Roll rate >20°/sec NOTE: Zero rudder authority available at 15° AOA.
NOTES	1. In takeoff/landing gains, the FLCS operates as a pitch rate command system until 10° AOA and a pitch rate/AOA command system above 10° AOA 2. +9g available until 15° AOA. Maximum g decreases as a function of AOA and airspeed.	1. In takeoff/landing gains, maximum roll rate is fixed at approximately one-half the maximum rate available in cruise gains, regardless of AOA, airspeed, or horizontal tail deflection. 2. Above 35° AOA, the yaw rate limiter cuts out stick roll commands and provides roll axis antispin control inputs.	1. Above 35° AOA, the yaw rate limiter provides yaw axis anti-spin control inputs. 2. Below -5 ° AOA and less than 170 knots, the yaw rate limiter provides antispin rudder inputs; pilot roll and rudder commands are cut out only when MPO is engaged. 3. Maximum deflection (30°) always available thru ARI and stability augmentation.

CENTRE AND RIGHT The CAT switch position determines the automatic control of pitch, roll and yaw axes. It is located on the left auxiliary panel. So too is the MPO – manual pitch override – switch, which can be held in the override (OVRD) position to enable the pilot to command full deflection to the horizontal tails in the event of a 'deep stall'. *(USAF)*

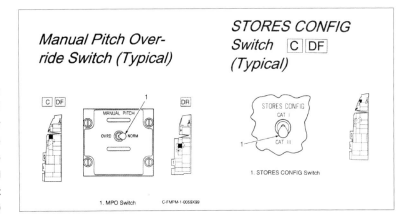

Manual Pitch Override Switch (Typical)

STORES CONFIG Switch C DF *(Typical)*

1. MPO Switch   C-FMPM-1-0059X99

1. STORES CONFIG Switch

RIGHT An F-16 unleashes an AIM-9M Sidewinder during a weapons trial. The Sidewinder has been continuously improved and updated, bearing little resemblance (in performance and probability of kill) to its forebears of the 1960s. Today's AIM-9X version provides high off-boresight capabilities. *(USAF)*

BELOW AND BOTTOM RIGHT The M56 high-explosive incendiary 20mm cannon round (left) was superseded for F-16 operations by the higher velocity PGU 28A/B 20mm HEI round, according to a 2007 US Air Force press release. The PGU-28's higher velocity provides greater stand-off distance for the pilot, and greater effect on impact. The rounds are loaded mechanically using a feed chute, although in this image they are actually being removed following a mission. *(USAF)*

BELOW The seeker head of the AIM-9M is cooled by liquid nitrogen to make it even more sensitive. Here, an F-16 pilot inspects a Sidewinder's rear fins – each containing a rolleron that stabilises the missile during its flight to the target. *(USAF)*

BELOW The AIM-120 AMRAAM (advanced medium-range air-to-air missile) gives the Viper the ability to engage multiple threats beyond visual range. Carried on either the outboard weapons pylons or the wingtip launchers, the missile is highly capable. *(USAF)*

WEAPONS: AIR-TO-AIR

The current standard air-to-air missiles (AAMs) carried on the F-16 are the AIM-9X Sidewinder and AIM-120B/C AMRAAM. The older AIM-7M Sparrow can be fired by modified F-16As and Cs, but its use is restricted to a handful of operators.

Several other missiles are also carried by F-16s. Pakistani Vipers are equipped with MATRA Magic 2s, Greek Vipers employ the Diehl BGT Defence IRIS-T short-range missile, while Israeli F-16s carry Rafael Python III and IV.

The most recent generation of short-range air-to-air missiles to enter service are also compatible with the F-16, and include the MBDA AIM-132 Advanced SRAAM (ASRAAM); Israel's Python V missile will equip the F-16 when it becomes operational.

For close-in engagements and ground strafing, the F-16 is still equipped with a gun, and the Falcon has an M61A1 20mm Vulcan cannon.

LEFT A number of versions of the AGM-65 Maverick can be carried by the F-16, including IR and EO (electro optical) varieties. The weapon is a fire-and-forget missile that can be employed against armour and hardened targets. *(USAF)*

WEAPONS: AIR-TO-GROUND

With the development of the Block 50/52 and the ASQ-213 HARM targeting system, the F-16 became an effective SEAD (suppression of enemy air defences) aircraft. When equipped with the Raytheon AGM-88 HARM, it is able to actively hunt SAMs rather than merely react to them. The integration of AAQ-28 Litening II and Sniper XR target pods to the Block 50 F-16 in 2005 has further enhanced the jet's SAM-hunting prowess.

The F-16 also has a wide range of precision-guided weapons to choose from, with the Paveway IV series of GBUs (glide bomb units) being the most recent family of laser-guided weapons cleared for operational use.

The AGM-65 Maverick remains a key F-16CM air-to-ground weapon. One of the latest weapons to be qualified for carriage on the F-16 is the satellite-guided family of weapons. Among these, the most ubiquitous

LEFT A series of images showing the nose [far left] and tail [left] sections of the joint direct attack munition (JDAM) guidance kit attached to a 500lb Mk82 bomb. The resulting weapon is designated the GBU-38 and can be released against targets designated by the Viper's HTS or target pod. (It can also be cued by the radar or released visually, albeit with lower accuracy.) Pairs of the weapon can be loaded on to the Viper [centre and bottom] on each of the main wing pylons. The GBU-38 is an excellent weapon for urban warfare, not only because of its accuracy, but also due to its relatively small blast effect footprint. *(USAF)*

ABOVE The GBU-12 is a laser-guided bomb created through the addition of a guidance and steering kit to a Mk82 500lb bomb. Spring-loaded fins pop out of the rear section, stabilising the weapon in flight [above right], while steering fins and laser seeker dominate the front guidance kit. *(USAF)*

is the JDAM (joint direct attack munition) series of bombs, which use a new tail section housing an inertial navigation system (INS) guidance unit, augmented by GPS and four clip-on fins to a basic bomb body. Other GPS weapons include the BLU-108B SFW

(sensor-fused weapon); AGM-154 JSOW (joint stand-off weapon); GBU-39 small diameter bomb (SDB); the wind-corrected munitions dispenser (WCMD); and the GBU-54 dual-mode laser JDAM.

Unguided Falcon weapons include Mk82 and 84 LDGP (low-drag general purpose) bombs, CBU-58H/B or -87 cluster bombs and rocket pods.

A wide range of non-US weapons are also available for the F-16, including: the Aérospatiale AS30L; Kongsberg Penguin; Rafael Popeye; Daimler-Benz DWS 39 gliding dispenser; and a range of indigenous Israeli weapons, including the Spice series of guided bombs.

BELOW Rockets are a relatively rare weapons load for the Viper. However, they are effective against soft targets and troops in the open, and marker versions (featuring white phosphorus warheads) can also be used to mark targets for attack by other aircraft. The blue warhead of this rocket indicates that it is an inert practice round. *(USAF)*

RIGHT The simplest of all the F-16's stores is this, the trusty Mk82 500lb LDGP (low-drag general purpose) 'dumb' bomb. *(USAF)*

Autopilot

The autopilot provides attitude hold, heading select and steering select in the roll axis, and attitude hold and altitude hold in the pitch axis. These modes are controlled by 'PITCH' and 'ROLL' switches on the MISC panel. The 'TRIM/AP DISC' switch on the 'MANUAL TRIM' panel disengages the autopilot. The paddle switch on the stick interrupts autopilot operation while the switch is depressed and held.

The 'PITCH' switch is a three-position switch that is solenoid-held in an engaged position and returns to 'A/P OFF' when any of a number of conditions are met (landing-gear handle down, for example).

Movement of the 'PITCH' switch out of 'A/P OFF' engages both the pitch and roll autopilot modes selected. The 'ROLL' switch is a three-position switch that enables one of the three roll autopilot modes whenever a pitch autopilot mode is selected.

The autopilot is fully engaged when the 'PITCH' switch is not in 'A/P OFF'. Autopilot options are selected by positioning the 'PITCH' switch ('ALT HOLD', 'A/P OFF' or 'ATT HOLD') and the 'ROLL' switch ('HDG SEL', 'ATT HOLD' or 'STRG SEL').

Stick trim is inoperative with the autopilot engaged. The manual trim is operable and may be used while the autopilot is engaged. However, due to the limited authority of the autopilot, engagement of any mode in other than a trimmed flight condition degrades autopilot performance.

The autopilot loop in the FLCC receives inputs from the INS and CADC (central air data computer) by means of the AMUX (analogue multiplexer) bus. A lack of data, inaccurate data, or degradation/failure of the AMUX disconnects the autopilot and activates the FLCS 'FAULT' caution light and the FLCS 'A/P FAIL PFL' message. If AoA is greater than 15° the autopilot disconnects and the FLCS 'FAULT' caution light and the FLCS 'A/P FAIL PFL' message activate.

Additionally, the FLCC monitors autopilot operation for a failure to maintain the selected mode and for prolonged engagement outside of autopilot attitude limits with no stick inputs. Detection of a failure results in activation of the FLCS 'FAULT' caution light and the FLCS 'A/P DEGR PFL' message.

Positioning the 'PITCH' switch to 'ALT HOLD' enables the FLCS to use CADC information to generate commands to the horizontal tails, which result in the aircraft maintaining a constant altitude. The FLCS limits the pitch command to +0.5g–+2g. Engagement of altitude hold at rates of climb or dive less than 2,000fpm selects an altitude within the pitch command g limits. Engagement above rates of 2,000fpm causes no unsafe manoeuvres; however, the engaged altitude may not be captured.

Control accuracy of ±100ft is provided to 40,000ft pressure altitude for normal cruise conditions. The altitude reference may be changed by depressing the paddle switch, changing altitude and releasing the paddle switch. 'ALT HOLD' in the transonic region may be erratic.

Positioning the 'PITCH' switch to 'ATT HOLD' allows an attitude signal from the INU (inertial navigation unit) to be used to maintain the selected pitch attitude.

Positioning the 'ROLL' switch to 'HDG SEL' allows the FLCS to use a signal from the HSI (horizontal situation indicator) to maintain the heading set on the HSI. Adjusting the HSI heading reference marker to the aircraft heading prior to engagement maintains the existing aircraft heading; otherwise when the autopilot is engaged with the 'ROLL' switch in 'HDG SEL', the aircraft turns to capture the heading indicated by the heading reference marker on the HSI. The roll command does not exceed a 30° bank angle or a 20°/second roll rate.

Positioning the 'ROLL' switch to 'ATT HOLD' routes an attitude signal from the INU to the FLCS which results in the aircraft maintaining the selected roll attitude.

Positioning the 'ROLL' switch to 'STRG SEL' allows the autopilot to steer the aircraft to the selected steerpoint using roll commands. The roll command does not exceed a 30° bank angle or a 20°/second roll rate.

None of the four modes above function when pitch angle exceeds ±60°.

A stick steering function is operable with the pitch and roll attitude hold modes, allowing the pilot to use the stick to increase or decrease the desired attitude, even after the autopilot is engaged. The altitude hold mode functions similarly.

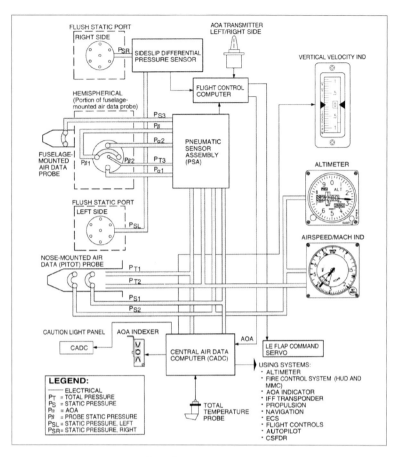

LEGEND:
— ELECTRICAL
P_T = TOTAL PRESSURE
P_S = STATIC PRESSURE
P_α = AOA
P_β = PROBE STATIC PRESSURE
P_{SL} = STATIC PRESSURE, LEFT
P_{SR} = STATIC PRESSURE, RIGHT

In flight the airflow direction is sensed by the conical AoA probes and the AoA ports of the fuselage-mounted air data probe. The AoA signals from all three sources are sent to the input selector/monitor of each of the four FLCC branches.

Air data system

The air data system uses probes and sensors to obtain static and total air pressures, AoA, sideslip, and air temperature inputs. These air data parameters are processed and supplied to various systems.

Proper AoA transmitter and fuselage air data probe operation is essential for safe flight operation. Interference from ice can result in erroneous data. This is very important, because false high AoA readings from two sources can cause the FLCC to command full nose-down pitch, which is impossible for the pilot to stop. With this in mind, probe covers protect the system from foreign objects and moisture intrusion on the ground, whilst heating elements in the probes and sensors can be turned on when icing conditions exist in flight.

Central air data computer (CADC)

The CADC receives total and static pressures, AoA, and total temperature inputs, then converts these into digital data. The data is then transmitted to those systems that

BELOW Once the CADC (central air data computer) has AoA information, it is displayed on both an indexer mounted to the left of the HUD and an indicator located on the central instrument panel. *(USAF)*

AoA system

The AoA (angle of attack) system consists of two AoA transmitters located on each side of the nose radome, AoA ports on the fuselage-mounted air data probe, a pneumatic sensor assembly (PSA), an AoA correction device in the CADC, an AoA indexer on the left of the HUD, and a vertical scale AoA indicator on the lower instrument panel.

AOA Displays

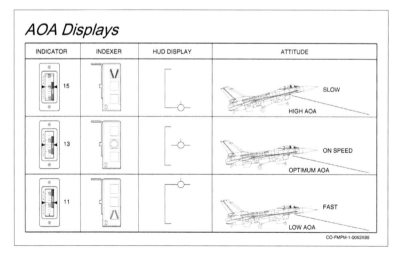

INDICATOR	INDEXER	HUD DISPLAY	ATTITUDE	
15	V			SLOW HIGH AOA
13				ON SPEED OPTIMUM AOA
11	Λ			FAST LOW AOA

CO-FMPM-1-0063X99

RIGHT A flow diagram showing the CADC's data inputs and outputs. Note how information from the computer is passed to a wide range of other systems and instruments, including the landing-gear configuration warning light. *(USAF)*

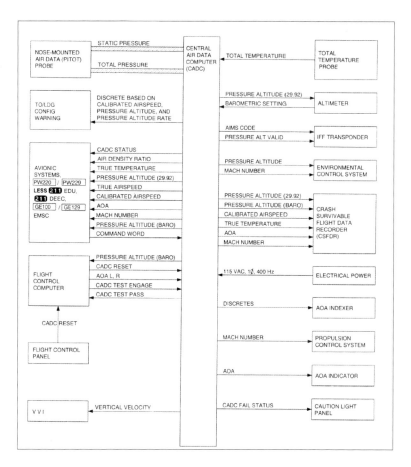

require it. The CADC has a continuous built-in check process.

Escape system and ACES II ejection seat

The canopy is a two-piece, plastic, bubble-type transparent enclosure consisting of two parts. The forward part is a single-piece transparency hinged at the aft end and unlatched, opened or closed/latched by an electrically operated actuator with a manual backup. The rear part is a smaller fixed transparency that fairs into the fuselage aft of the seat. The canopy may be jettisoned by internal controls for in-flight or ground escape and by external controls for ground rescue.

An inflatable pressurisation seal on the cockpit sill mates with the edge of the front canopy, while a non-inflatable rubber seal on the canopy prevents the entry of water when the cockpit is not pressurised.

The canopy provides some bird-strike protection: on the centreline, at approximately eye level, there may be enough flex that the HUD combiner glass will shatter. Deflection of the canopy in the area of the pilot's helmet has been observed to be 1–2in during bird-strike tests. These tests involved the use of a 4lb bird at 350kt or 550kt, and were required to ensure that the canopy did not deflect more than 2¼in in the area of the pilot's helmet. However, F-16 pilots do need to consider their seat height adjustment, especially while flying at lower altitudes with helmet-mounted equipment. Impacts off-centre of the forward canopy cause a less severe flex.

An internal manual hand-crank can be used to open or close the canopy, although the canopy control switch (below the canopy sill, just forward of the throttle) is normally used. Ground crew can jettison the canopy by pulling either of the two external canopy D-handles, located on either side of the fuselage.

BELOW Canopy controls for the F-16 include those for normal and emergency operations. Whilst the canopy is jettisoned automatically when the ACES II is fired, ground egress without ejection can be accomplished by either opening the canopy manually, or, when time is critical, jettisoning it. *(USAF)*

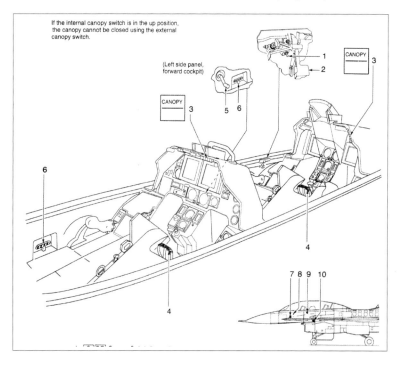

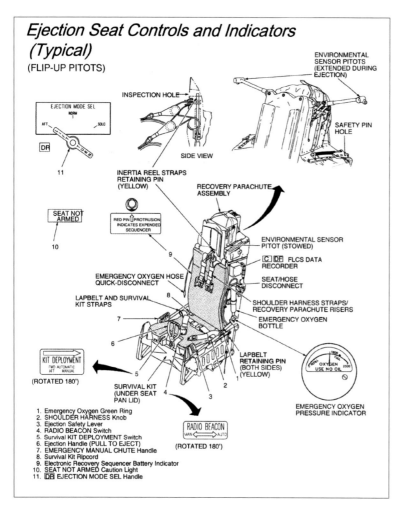

Ejection Seat Controls and Indicators (Typical)
(FLIP-UP PITOTS)

ENVIRONMENTAL SENSOR PITOTS (EXTENDED DURING EJECTION)

SAFETY PIN HOLE

EJECTION MODE SEL
NORM
AFT — SOLO
DR
11

INSPECTION HOLE

SIDE VIEW

INERTIA REEL STRAPS RETAINING PIN (YELLOW)

RECOVERY PARACHUTE ASSEMBLY

SEAT NOT ARMED
10

RED PIN PROTRUSION INDICATES EXPENDED SEQUENCER

ENVIRONMENTAL SENSOR PITOT (STOWED)

C DF FLCS DATA RECORDER

SEAT/HOSE DISCONNECT

EMERGENCY OXYGEN HOSE QUICK-DISCONNECT

LAPBELT AND SURVIVAL KIT STRAPS
7

SHOULDER HARNESS STRAPS/ RECOVERY PARACHUTE RISERS

EMERGENCY OXYGEN BOTTLE

6

KIT DEPLOYMENT
FWD AUTOMATIC AFT MANUAL
(ROTATED 180')

5

SURVIVAL KIT (UNDER SEAT PAN LID)
4

LAPBELT RETAINING PIN (BOTH SIDES) (YELLOW)

2

3

OXYGEN USE NO OIL

EMERGENCY OXYGEN PRESSURE INDICATOR

RADIO BEACON
MAN — AUTO
(ROTATED 180')

1. Emergency Oxygen Green Ring
2. SHOULDER HARNESS Knob
3. Ejection Safety Lever
4. RADIO BEACON Switch
5. Survival KIT DEPLOYMENT Switch
6. Ejection Handle (PULL TO EJECT)
7. EMERGENCY MANUAL CHUTE Handle
8. Survival Kit Ripcord
9. Electronic Recovery Sequencer Battery Indicator
10. SEAT NOT ARMED Caution Light
11. DR EJECTION MODE SEL Handle

ABOVE The ACES II ejection seat is a successful design that is trusted by those whose lives depend upon it. Note how this diagram shows the version of the seat that has environmental sensor pitots that extend upon ejection – another version of the seat used by the F-16 features these sensors in a fixed extended position. *(USAF)*

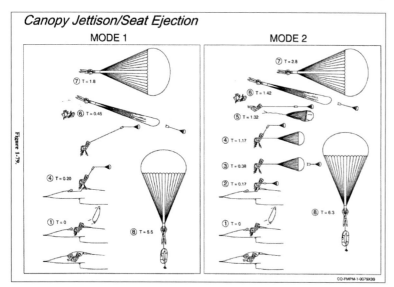

Canopy Jettison/Seat Ejection

MODE 1

⑦ T = 1.8
⑥ T = 0.45
④ T = 0.20
① T = 0

MODE 2

⑦ T = 2.8
⑥ T = 1.42
⑤ T = 1.32
④ T = 1.17
③ T = 0.38
② T = 0.17
① T = 0
⑧ T = 6.3
⑧ T = 5.5

Figure 1-79.

CO-FMPM-1-0079X99

Pulling the ejection handle ('PULL TO EJECT'), located on the front of the ejection seat, initiates the canopy jettison sequence followed by the seat ejection sequence.

The ACES II ejection seat is a fully automatic emergency escape system. One of three ejection modes is automatically selected. Mode 1 is a low-airspeed, low-altitude mode during which the recovery parachute assembly is deployed almost immediately after the ejection seat departs the aircraft. Mode 2 is an intermediate-airspeed, low-altitude mode during which a drogue chute is first deployed to slow the ejection seat, followed by the deployment of the recovery parachute assembly. Mode 3 is a high-airspeed/high-altitude mode in which the sequence of events is the same as mode 2, except that automatic pilot/seat separation and deployment of the recovery parachute assembly are delayed until safe airspeed and altitude are reached.

The ejection handle ('PULL TO EJECT') is sized for one-handed or two-handed operation, and requires a pull of 40–50lb to activate. The handle remains attached to the seat by a wire cable after activation.

A shoulder harness knob unlocks the inertia reel when in the aft position and locks it when in the forward position. In the same way as the seatbelt in an automobile is designed, if high longitudinal deceleration force or high shoulder harness strap velocities are encountered, the inertia reel automatically locks.

Oxygen system

The oxygen system consists of an on-board oxygen generating system (OBOGS), a backup oxygen supply (BOS), and a regulated $50in^3$ emergency oxygen supply (EOS) bottle.

The OBOGS is a molecular sieve/concentrator that uses an absorption filtering process to remove nitrogen from the ECS

LEFT The ACES II is a fully automatic ejection seat, meaning that the pilot simply needs to pull the handle to initiate the ejection sequence. Once that sequence commences, the pilot is required to do nothing else until his parachute canopy is inflated. *(USAF)*

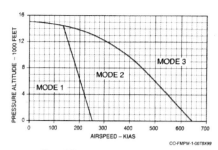

Ejection Mode Envelopes

Figure 1-77.

Ejection Sequence Times

		TIME (SECONDS)	
EVENT		Mode 1	Mode 2
1. Catapult Initiation		0.0	0.0
2. Drogue Gun Fired		NA	0.17
3. Drogue Chute Inflated		NA	0.38
4. Parachute Fired		0.20	1.17
5. Seat/Drogue Separation		NA	1.32
6. Pilot/Seat Separation		0.45	1.42
7. Recovery Parachute Inflated		1.8	2.8
8. Survival Kit Deployed		5.5	6.3

NOTE
- In mode 3, events after drogue deployment are delayed until within mode 2 envelope. Recovery parachute deploys 1 second after entering mode 2 envelope.
- ▣ Times in the aft/forward sequence increase to include a 0.33-second delay for the rear seat and a 0.73-second delay for the forward seat. In SOLO, the forward seat is delayed 0.33 second.
- Canopy jettison time varies from 0.75 second at 0 KIAS to 0.13 second at 600 KIAS. Ejection begins when canopy jettison initiates seat lanyards.

(environmental control system), which supplies bleed air and provides oxygen-rich breathing gas. The system is capable of producing 95% oxygen.

For normal operation, the concentrator provides oxygen to a compatible diluter demand-breathing regulator. The system pressure to the regulator can be read on a pressure gauge located on the face of the regulator. The normal operating pressure range is 25–40psi.

If the OBOGS is not producing oxygen, the system will revert to stored oxygen in the sieve-filled plenum, called the backup oxygen supply (BOS). The BOS will last for a period of 3–5 minutes depending on altitude and breathing rate. Resupplying a depleted BOS requires up to 10 minutes. After the BOS has been expended and system pressure falls below 5psi, the 'OXY LOW' warning light, located on the right glare shield, illuminates.

The EOS consists of a high pressure bottle mounted on the left side of the ejection seat. The EOS is a regulated oxygen supply that will last for 8–12 minutes depending on altitude and breathing rate. The hose is routed to the

right side of the seat. The system is activated automatically upon ejection, or manually by pulling the emergency oxygen green ring located on the left aft side of the seat.

A pressure breathing for g (PBG) mode on the cockpit oxygen regulator provides pressure breathing above 4g to enhance g tolerance and reduce fatigue. Air pressure from the anti-g valve is used by the oxygen regulator to control the amount of pressurisation supplied to the Combat Edge oxygen mask, helmet and (where used) the vest.

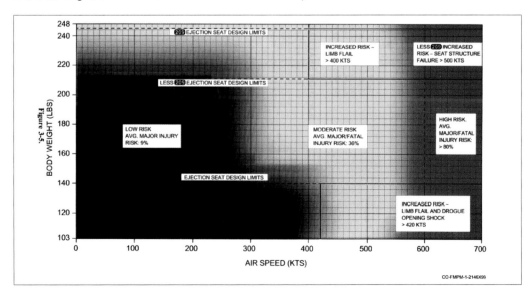

LEFT A colour-coded chart showing body-weight plotted against airspeed for the ACES II. The red zone to the right starts at about 500 knots and shows a high risk of seat structure failure, fatal injury and flailing limb injury during ejection. (USAF)

Upfront Controls (Typical)

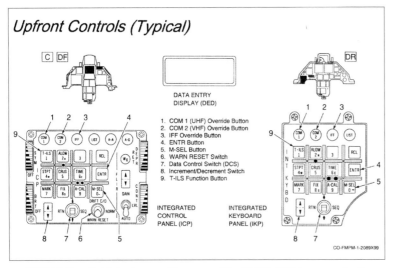

DATA ENTRY DISPLAY (DED)

1. COM 1 (UHF) Override Button
2. COM 2 (VHF) Override Button
3. IFF Override Button
4. ENTR Button
5. M-SEL Button
6. WARN RESET Switch
7. Data Control Switch (DCS)
8. Increment/Decrement Switch
9. T-ILS Function Button

INTEGRATED CONTROL PANEL (ICP)

INTEGRATED KEYBOARD PANEL (IKP)

CO-FMPM-1-2089X99

ABOVE The pilot interfaces with many of the F-16's CNI (communications, navigation and IFF) systems by means of the upfront control panel (front-seat UFC shown on left, back-seat on right). The panel features an array of mode and alpha-numeric keys that are logically arranged to allow quick access to the aircraft's systems. In the front seat, the UFC is mounted directly below the HUD. *(USAF)*

Communications, navigation and IFF (identification friend or foe) system

C ockpit controls for CNI equipment are divided between the consoles and the up-front controls located on the instrument panel. Controls for less frequently used functions, such as power and audio volume, and essential functions, such as communications backup and guard, are located on console panels. Controls for frequently used functions are located on the up-front controls to permit head-up control during flight.

BELOW The pilot fault display (PFD), mounted on the right auxiliary control panel, provides a list of faults with aircraft systems that the pilot can scroll through. *(USAF)*

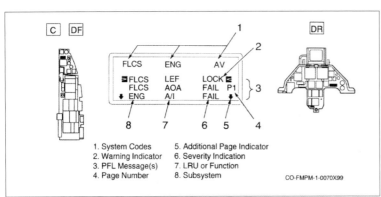

1. System Codes
2. Warning Indicator
3. PFL Message(s)
4. Page Number
5. Additional Page Indicator
6. Severity Indication
7. LRU or Function
8. Subsystem

CO-FMPM-1-0070X99

The up-front controls provide a simplified, centralised, head-up means of controlling the most frequently used functions of the communications system, navigation systems and IFF. The up-front controls consist of the data entry display (DED), the integrated control panel (ICP) and the integrated keyboard panel (IKP).

The up-front controls are powered by emergency dc bus No2 and emergency ac bus No2. The DED is an integral part of the up-front controls and provides a visual display of the switch actions made via up-front controls. The primary readouts of communication, navigation, and IFF systems are included in the page selections available for display on the DED. Channel, frequency, mode and code selections of UHF, VHF, TACAN, ILS and IFF are presented when the appropriate page is selected. INS present position and steerpoint data may be selected for display on the DED.

The INS is a prime sensor for aircraft velocity, attitude and heading and is a source of navigation information. In conjunction with the up-front controls, GPS, CADC and MMC (modular mission computer), the INS provides: present position with update and storage capability; current winds; ground speed and drift angle; great circle course computation; and instantaneous and maximum g data for display in the HUD.

The TACAN (tactical air navigation) system provides continuous bearing and distance information from any selected TACAN station within a line of sight distance up to approximately 390 miles, depending upon terrain and aircraft altitude. Only distance information is presented when a DME navigational aid is selected. There are 252 channels available for selection. The TACAN bearing, selected course, range and course deviation information are displayed on the HSI, as determined by the HSI mode button. TACAN is not available if the MMC is either failed or operating in a degraded mode. The DED displays information about four TACAN items: operating mode, channel number, band and TACAN station identifier.

The AIFF (Advanced IFF) system provides selective identification feature (SIF), automatic altitude reporting and encrypted mode 4 IFF. Normal operation is possible in any of six modes:

Mode 1 – Security identity.
Mode 2 – Personal identity.
Mode 3/A – Traffic identity.
Mode 4 – Encrypted identity.
Mode C – Altitude reporting.
Mode S – Air traffic control data link (includes mode 3/A and C functions).

The AIFF transponder only transmits coded replies to correctly coded interrogations. It provides for selection of manual or automatic operation. If the automatic function is selected, IFF operation can be programmed to turn modes on or off when a certain position is reached; change codes when a specific time is reached; turn modes on or off when a certain position is reached; and change codes when a specific time is reached. IFF modes may be selected and programmed for use in groups.

The AIFF interrogator provides selective interrogation of IFF systems along a line of sight (LOS) or within a specific scan area. For a specific scan area, the interrogator may be selectively coupled or decoupled with the fire control radar.

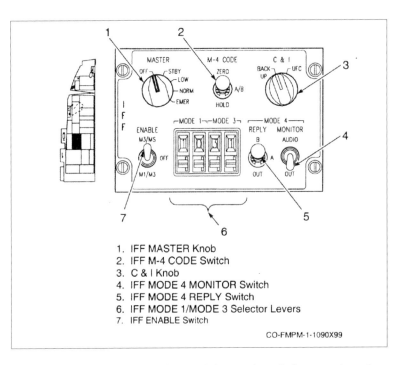

1. IFF MASTER Knob
2. IFF M-4 CODE Switch
3. C & I Knob
4. IFF MODE 4 MONITOR Switch
5. IFF MODE 4 REPLY Switch
6. IFF MODE 1/MODE 3 Selector Levers
7. IFF ENABLE Switch

CO-FMPM-1-1090X99

ABOVE The IFF panel is located on the left console and allows mode and code changes to be made quickly. It also allows the military's encrypted Mode 4 settings to be entered and stored. *(USAF)*

BELOW The F-16CM has a very intuitive set of instrumented landing system (ILS) HUD cues that the pilot can follow. Once on final approach, he simply has to manoeuvre so that his flight-path marker remains overlaid on to a circular steering symbol. *(USAF)*

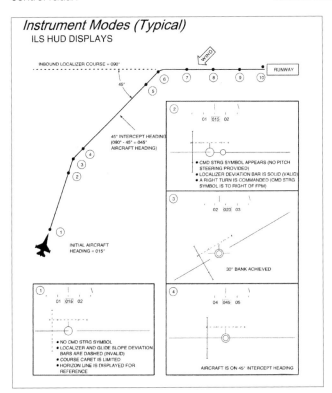

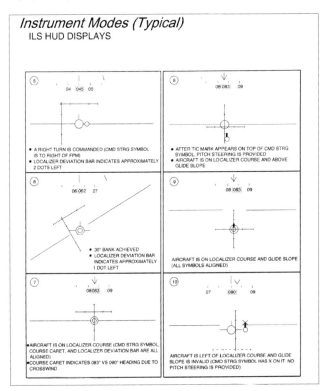

Chapter Six

The Pratt & Whitney F100-PW-229 and General Electric F110-GE-129 engines

The Block 40 and Block 50 F-16CM are powered by the F100-PW-220, -220E (both 25,000lb thrust) or -229 afterburning turbofan engine (29,000lb thrust), while the Block 42 and Block 52 are powered by the F110-GE-100 (28,000lb thrust on the large-inlet aircraft, 25,000lb on the small-inlet aircraft) or the -129 afterburning turbofan engine (29,500lb thrust).

OPPOSITE The flat exhaust petals, or 'turkey feathers', of this F-16C Block 25, identify it as being equipped with a Pratt & Whitney engine. *(USAF)*

101

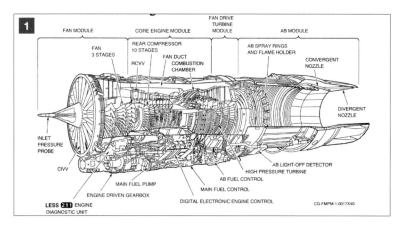

LEFT The four engine versions used to power the F-16CM are the F100-PW-220 [1] and PW-229 [2], and the F110-GE-100 [3] and GE-129 [4]. The differences between the four motors are illustrated in these cutaway diagrams. *(USAF)*

For the sake of brevity, the improved performance engines – PW-229 and GE-129 are discussed below. Whilst each engine has its own set of limitations and idiosyncrasies, the commonality between them is great. Where a component or system is particular to one engine or the other, this will be indicated in parentheses as either (GE-129) or (PW-229).

F110-GE-129

General Electric describes the F110 series of engines as 'sharing 81% parts commonality with the F110-100', and notes that 'the -129 offers significant mission advantages throughout the F-16 flight envelope'. Its marketing literature further adds that 75% of the most modern F-16s are powered by the F110 family of engine.

Fan stages: 3
Compressor stages: 9
High-pressure turbine/low-pressure turbine stages: ½
Maximum diameter: 46.5in
Length: 182.3in
Overall pressure ratio at maximum power: 30:7

F100-PW-229

Of the PW-229, Pratt & Whitney states that 'the most significant accomplishment of the F100 family is its extraordinary flight safety record. The F100-PW-229 engine has the best safety record of any increased-performance engine.'

Fan stages: 3
Compressor stages: 10
Weight: 3,740lb
Length: 191in
Inlet diameter: 34.8in
Maximum diameter: 46.5in
Bypass ratio: 0.36
Overall pressure ratio: 32:1

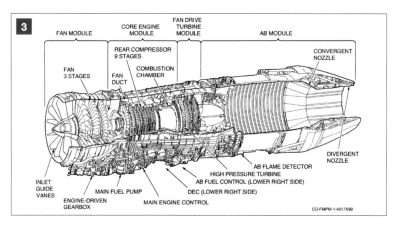

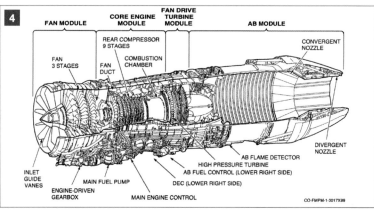

Engine fuel control system

The engine fuel/control system delivers the required fuel to the engine for combustion and for use by the control system for scheduling the engine variable geometry.

The control system is primarily composed of three major components: digital electronic control (DEC) (GE-129) or digital electronic engine control (DEEC) (PW-229); the afterburner fuel control (AFC); and the main engine control (MEC) (GE-129) or main fuel control (MFC) (PW-229).

Both engines have two pilot selectable modes of operation: primary (PRI) and secondary (SEC). The GE-129 also has a hybrid mode that bridges the gap between PRI and SEC. However, this is not pilot-selectable: it is

automatically entered when the DEC registers certain failure conditions.

The DEC/DEEC is the critical component of the primary engine control. It is an engine-mounted, fuel-cooled solid-state digital

LEFT The inlet duct routes air to the fan module. Here, it first meets the inlet guide vanes (GE engines) or compressor inlet variable vane control (PW engines), before driving three fan stages. The inlet is large enough to crawl inside for inspection. The mouth size – 'big' or 'small' – of the inlet is determined by the engine version installed. (USAF)

RIGHT The afterburner section comprises divergent nozzles ([1] nearest the camera), convergent nozzles ([2] photographed facing aft) and AB spray rings and flameholder ([3] also photographed facing aft). The spray rings inject raw fuel into the hot exhaust, while the nozzles adjust to control pressure inside the AB section itself. *(USAF)*

computer that controls both the main engine and the afterburner (AB).

The afterburner fuel control is a fuel-operated electrohydromechanical control that regulates fuel flow to the AB in conjunction with the DEC/DEEC. During primary operation, the AB fuel control receives fuel from the AB fuel pump and electrical commands from the DEEC/DEC. It provides AB ignition, AB segment sequencing, and fuel flow to the AB segments. During SEC control, AB fuel flow is inhibited.

The MEC and MFC are fuel-operated hydromechanical controls that provide engine control functions in both PRI and SEC modes. During PRI control, the MFC receives throttle

inputs, fuel from the main fuel pump, and electrical commands from the DEEC. It controls main ignition, start bleed strap position, main engine fuel flow, and rear compressor variable vane (RCVV) position.

The MFC also provides actuation pressure to the compressor inlet variable vane (CIVV) control, the convergent exhaust nozzle control (CENC), and both the AB fuel control and AB pump controller.

Primary engine operation

PRI provides unrestricted engine operation throughout the entire flight envelope. Control provided by the DEC/DEEC during PRI operation includes: fan speed control, fuel flow scheduling, turbine blade/inlet temperature limiting, exhaust nozzle control, ignition logic for starting and automatic relight sequencing in both the engine and AB, and engine overspeed protection (113% rpm overspeed fuel shut-off valve for the GE-129).

The exhaust nozzle is controlled by signals from the DEC/DEEC to the engine hydraulic pump, which positions four nozzle actuators in order to maintain fan stall margin while providing the requested level of thrust. Meanwhile, fan inlet guide vane (IGV) positioning is controlled by the DEC/DEEC in accordance with a pre-programmed schedule.

During transonic and supersonic flight, with the throttle retarded below MIL, the DEC limits minimum engine operation as a function of Mach number from the central air data computer (CADC) to prevent inlet buzz and possible engine stall. When retarding the throttle to 'IDLE' above 1.4 Mach, rpm may decrease up to 15% from MIL (military power) rpm. The rpm then decreases with Mach number until approximately 1.1 Mach, at which time the engine decelerates to normal flight idle rpm.

Secondary engine operation

SEC is activated by either manually placing the 'ENG CONT' switch to 'SEC' or as a result of automatic transfer when the DEC detects certain failures. In 'SEC', the MEC/MFC provides fuel flow scheduling and metering and other functions, but it closes the exhaust nozzle and inhibits AB operation. With the exhaust nozzle shut, MIL thrust is 70–95% of that provided in PRI mode and 'IDLE' thrust is higher.

Engine components and accessories

Engine fuel boost pump (GE-129)

A gearbox-mounted engine fuel boost pump provides pressurised fuel to the main fuel pump and AB fuel pump.

Main fuel pump

A gear-type main fuel pump receives pressurised fuel from the engine fuel boost pump and supplies extra pressure to the MEC on the GE-129. On the PW-229, it provides pressurised fuel to the MFC, and boosts pressure to the AB fuel pump.

Afterburner (AB) fuel pump

The AB fuel pump is mounted on the engine gearbox and receives fuel from the engine fuel boost pump.

Inlet guide vanes (GE-129)

Each IGV is an airfoil that is divided into two sections. The forward portion of the inlet guide vane is fixed, which provides structural support. The aft portion of the inlet guide vane is a variable angle flap, which controls the angle at which air enters the fan. This both improves fan efficiency and increases the stall margin.

Variable stator vanes (GE-129)

The compressor VSV system controls the angle of the core inlet guide vanes and the first three stages of core variable stator vanes. Positioning is a function of engine rpm. By varying the vane position, the system automatically changes the effective angle at which the airflow enters the compressor rotor blades, thereby maintaining satisfactory airflow and optimum compressor performance throughout the entire flight envelope. For increased stall protection the VSVs are reset slightly closed from their normal position after a throttle snap to 'IDLE'. The reset position is maintained for two minutes, after which the VSVs return to their normal schedule, resulting in an rpm drop of approximately 2%.

Inlet variable vane controller (PW-229)

Instead of the GE-129's IGVs and VSVs, the PW-229 features a compressor inlet variable vane controller. In PRI mode, the controller positions the compressor vanes using MFC fuel pressure in response to an electrical signal from the DEEC. In SEC, the controller fixes the vanes in a cambered position.

Rear compressor variable vanes (PW-229)

The first three stages of the rear compressor are equipped with variable geometry vanes. In PRI, these are controlled by the DEEC and are positioned using pressurised fuel from the main fuel pump. In SEC, they are positioned by a hydromechanical control in the MFC as a direct function of throttle position.

Compressor bleed air (GE-129)

Bleed air is extracted from two separate stages in the compressor for engine and airframe use. Low-pressure (fifth stage) air is used for turbine cooling and the engine anti-ice system. Air for airframe use is taken from both the low- and high-pressure (ninth stage) compressor sections. Low-pressure bleed air is used for the ECS unless the pressure is insufficient, in which case high-pressure bleed air is used. High-pressure bleed air is used for the nacelle ejectors, and also to power the EPU.

Compressor bleed air (PW-229)

Low-pressure bleed air is directed from the bleed strap into the fan duct to increase the compressor stall margin during starting. Pressurised fuel from the main fuel pump is used to drive the start bleed actuator. The bleed valve is scheduled as a function of engine rpm by the DEEC when starting in PRI and as a function of time and engine inlet pressure in SEC.

High-pressure bleed air is supplied to the EPU and engine nacelle ejectors. It is also used for engine inlet anti-icing, to drive the AB fuel pump. Either low-pressure or high-pressure air is provided to the ECS, depending on engine bleed pressure levels.

Pressurisation and dump valve (PW-229)

A pressurisation and dump valve is located in the engine fuel manifold line between the fuel/oil cooler and fuel nozzles. It provides a minimum fuel pressure for MFC operation at low rpm; the dump port is capped so that fuel is not drained from the engine fuel manifold when the throttle is moved to 'OFF'.

Exhaust nozzle

The exhaust nozzle is a variable area convergent/divergent semi-floating type, with mechanically linked primary and secondary flaps and seals. Nozzle area modulation is accomplished by hydraulic actuators operated by the engine hydraulic pump, using engine oil as hydraulic fluid. The actuators respond to electrical commands from the DEC/DEEC.

The primary functions of the nozzle system are to maintain fan stall margin by varying the nozzle area and to control the engine thrust for optimum performance through the entire flight envelope.

The PW-229 also features a convergent nozzle control, which has multiple functions but can be summarised as using DEEC commands to schedule nozzle position according to a range of factors.

AB flame detector (GE-110) and light-off detector (PW-229)

The AB flame detector provides AB light/no-light information to the DEC for use in AB sequencing and auto-relight functions during PRI operation. The light-off detector in the PW-229 accomplishes much the same.

Engine monitoring system (GE-110) and engine diagnostic unit (PW-229)

The EMS is operative in all engine control modes and is designed to perform engine diagnostics and store engine fault data for post-flight analysis. The EMS consists of two primary components: the digital electronic control (DEC), which is mounted on the engine, and an engine monitoring system, which is located in the leading-edge flap drive bay in the forward fuselage section.

Event data is automatically captured when a fault is detected, but the pilot can manually store data at any time by placing the 'AB RESET' switch to 'ENG DATA'.

The PW-229's EDU operates in conjunction with the DEEC, and provides almost identical functionality to GE's EMS.

Engine oil system

The engine is equipped with a self-contained, dry-sump, full-pressure lubrication system, which provides filtered oil for lubricating and cooling the engine main shaft bearings, oil seals, gearboxes and accessories. It also provides oil to the engine hydraulic pump for nozzle actuation.

'Fuel hot' caution light

Located on the caution light panel, the 'FUEL HOT' caption illuminates when the temperature of the GE-129's oil becomes excessive (exceeds 300°F) or when excessive fuel temperature is observed. The caution light is inoperative on aircraft equipped with the PW-229.

Engine and accessory drive gearboxes

The GE-129's engine gearbox drives the main fuel pump, the engine fuel boost pump, the AB fuel pump, the engine/scavenge pump, the engine alternator, the engine hydraulic pump, the MEC, and a shaft that powers the accessory drive gearbox (ADG). On the PW-229 it drives the main fuel pump, the oil pump assembly, the engine alternator, and the shaft that powers the ADG.

The ADG powers the main generator through a constant-speed drive (CSD), system A and B hydraulic pumps, standby generator, and the FLCS PMG. The JFS is also mounted on the ADG.

Jet fuel starter

The JFS is a gas turbine. It operates on aircraft fuel and drives the engine through the ADG. The JFS is connected by a clutch to the ADG and only provides torque when required to maintain engine rpm. It receives fuel at all times regardless of the 'FUEL MASTER' switch position, and is started by power from two brake/JFS accumulators used either singly or

RIGHT Engine accessory drive and gearboxes (PW left, GE right) connect to the engine to power other aircraft systems. The PW and GE gearboxes differ only slightly in their design. *(USAF)*

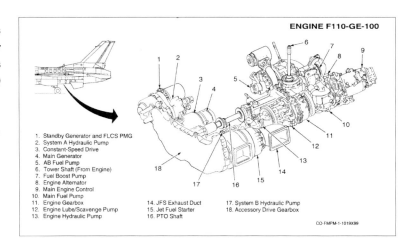

ENGINE F110-GE-100

1. Standby Generator and FLCS PMG
2. System A Hydraulic Pump
3. Constant-Speed Drive
4. Main Generator
5. AB Fuel Pump
6. Tower Shaft (From Engine)
7. Fuel Boost Pump
8. Engine Alternator
9. Main Engine Control
10. Main Fuel Pump
11. Engine Gearbox
12. Engine Lube/Scavenge Pump
13. Engine Hydraulic Pump
14. JFS Exhaust Duct
15. Jet Fuel Starter
16. PTO Shaft
17. System B Hydraulic Pump
18. Accessory Drive Gearbox

CO-FMPM-1-1019X99

together. The JFS is used to start the engine on the ground and to assist in engine air-start.

During a ground engine start, the brake/JFS accumulators begin to recharge after the engine accelerates through 12% rpm. As the engine accelerates through approximately 55% rpm, a sensor causes the JFS to shut down automatically.

Fuel system

The fuel system is divided into seven functional categories. These are the fuel tank system, fuel transfer system, fuel tank vent and pressurisation system, engine fuel supply system, fuel quantity/fuel level sensing system, fuel tank explosion suppression system, and refuelling/defuelling system.

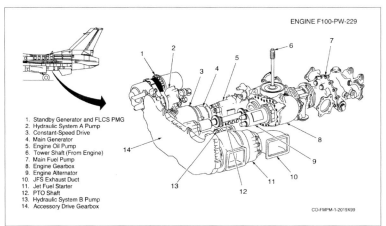

ENGINE F100-PW-229

1. Standby Generator and FLCS PMG
2. Hydraulic System A Pump
3. Constant-Speed Drive
4. Main Generator
5. Engine Oil Pump
6. Tower Shaft (From Engine)
7. Main Fuel Pump
8. Engine Gearbox
9. Engine Alternator
10. JFS Exhaust Duct
11. Jet Fuel Starter
12. PTO Shaft
13. Hydraulic System B Pump
14. Accessory Drive Gearbox

CO-FMPM-1-2019X99

BELOW A typical fuel system schematic for the **F-16CM.** *(USAF)*

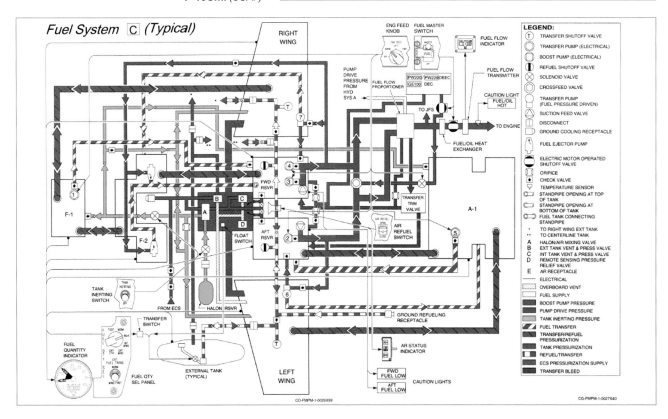

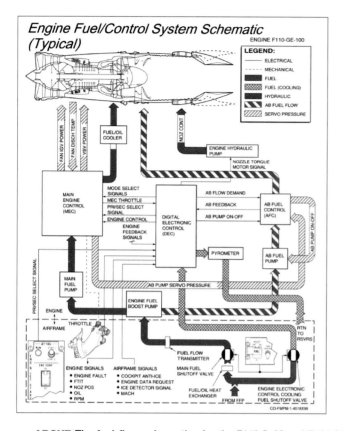

Engine Fuel/Control System Schematic (Typical)

ENGINE F110-GE-100

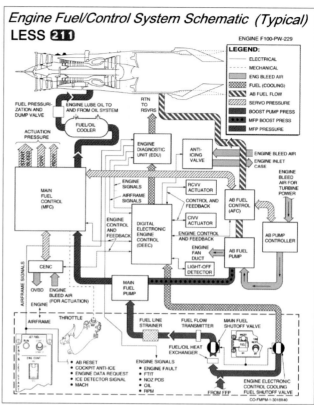

Engine Fuel/Control System Schematic (Typical)

LESS 211

ENGINE F100-PW-229

ABOVE The fuel-flow schematics for the F110 (left) and F100 (right) show small but important differences between the flow of fuel to and from the engine and aircraft systems. *(USAF)*

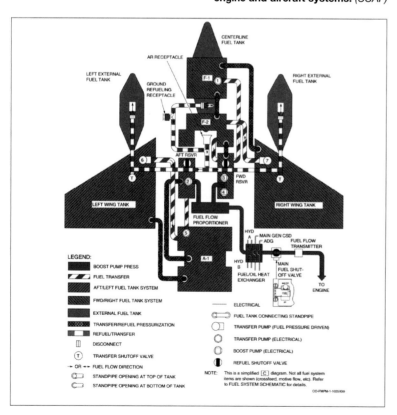

The F-16CM has seven internal fuel tanks located in the fuselage and wings that are integral to the structure. Five of the internal tanks are storage tanks: the left and right wing tanks, two forward fuselage tanks (F-1 and F-2), and the aft fuselage tank (A-1). The two internal reservoir tanks (forward and aft) supply fuel directly to the engine. On the F-16D, the F-1 fuel tank is reduced in size to allow room for the rear cockpit.

Fuel is transferred by two independent methods. The primary method provides a siphoning action through standpipes connecting the fuel tanks. Siphoning action depends on the absence of air in the bays receiving fuel. Air ejectors in each reservoir tank automatically expel air. In case of failure of the siphoning system, powered fuel pumps work continually to pump fuel from the internal tanks to the reservoirs. The powered transfer system also scavenges tanks to minimise unusable fuel by using electrically driven pumps and pumps

LEFT The F-16CM's fuel tank ordering is broken into the following main areas: wing tanks, external fuel tanks, aft system and forward system. *(USAF)*

powered by bleed fuel pressure from the engine manifold. Both methods operate simultaneously and independently to transfer fuel through the system.

The transfer system is divided into two separate tank systems, the forward and the aft. The forward system consists of the right external tank (if installed), right internal wing tank, F-1, F-2, and the forward reservoir. The aft system consists of the left external tank (if installed), left internal wing tank, A-1, and the aft reservoir. If a centreline tank is installed it is considered to be part of both forward and aft systems. The wing external tanks empty into the respective internal wing tanks.

Fuel flows from the internal wing tanks to the fuselage tanks and then to the forward and aft reservoirs. Fuel is pumped to the engine from the reservoirs. To automatically maintain the centre of gravity, fuel is transferred through the forward and aft systems simultaneously.

If external tanks are installed, air pressure transfers fuel to the internal wing tanks. If the 'EXT FUEL TRANS' switch is in 'NORM', the sequence of fuel flow is from the centreline tank to the internal wing tanks. After the centreline tank empties, each external wing tank flows to its respective internal wing tank. The external tank fuel transfer valve in each internal wing tank shuts off fuel to prevent overfilling the internal tanks. If one of these valves fails, a float switch senses fuel and shuts off all external tank fuel transfer before fuel flows overboard. By placing the 'EXT FUEL TRANS' switch to 'WING FIRST', the external wing tanks empty before the centreline tank, and the float switch does not prevent fuel from spilling overboard if a transfer valve fails.

The automatic forward fuel transfer system prevents undesirable aft centre of gravity. The automatic forward fuel transfer system operates only when the 'FUEL QTY SEL' knob is in 'NORM' and the total forward fuselage fuel quantity indication is less than 2,800lb. Forward fuel transfer starts when the forward heavy fuel differential drops below 300lb and stops when the forward heavy fuel differential reaches 450lb. In the F-16D, forward fuel transfer starts when the aft heavy fuel differential exceeds 900lb and stops when the aft heavy fuel differential reaches 750lb. This system does not correct a forward fuel imbalance since it only transfers fuel from aft to forward.

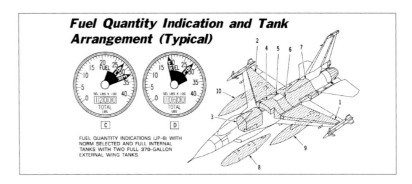

ABOVE A single fuel quantity indicator (here showing that in the C- and D-model Viper) allows the pilot to see total fuel quantity at a glance. The totaliser (displayed as digits) provides a total figure, while the pointer hands show forward/aft and right/left quantities. *(USAF)*

BELOW A mode selector allows the pilot to cycle between viewing fore, aft, left and right, external and total quantities. He may also cycle between internal and external fuel feed, although the external tanks are usually used first to lighten them and allow better combat manoeuvring without overstressing them. *(USAF)*

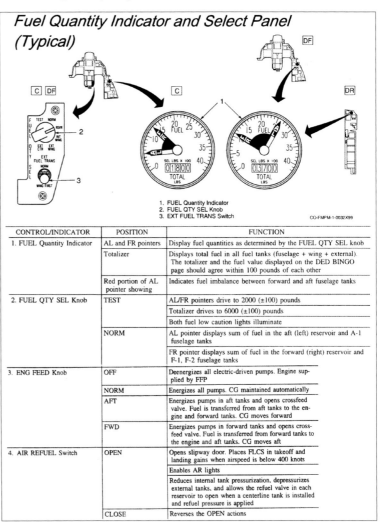

CONTROL/INDICATOR	POSITION	FUNCTION
1. FUEL Quantity Indicator	AL and FR pointers	Display fuel quantities as determined by the FUEL QTY SEL knob
	Totalizer	Displays total fuel in all fuel tanks (fuselage + wing + external). The totalizer and the fuel value displayed on the DED BINGO page should agree within 100 pounds of each other
	Red portion of AL pointer showing	Indicates fuel imbalance between forward and aft fuselage tanks
2. FUEL QTY SEL Knob	TEST	AL/FR pointers drive to 2000 (±100) pounds
		Totalizer drives to 6000 (±100) pounds
		Both fuel low caution lights illuminate
	NORM	AL pointer displays sum of fuel in the aft (left) reservoir and A-1 fuselage tank
		FR pointer displays sum of fuel in the forward (right) reservoir and F-1, F-2 fuselage tanks
3. ENG FEED Knob	OFF	Deenergizes all electric-driven pumps. Engine supplied by FFP
	NORM	Energizes all pumps. CG maintained automatically
	AFT	Energizes pumps in aft tanks and opens crossfeed valve. Fuel is transferred from aft tanks to the engine and forward tanks. CG moves forward
	FWD	Energizes pumps in forward tanks and opens crossfeed valve. Fuel is transferred from forward tanks to the engine and aft tanks. CG moves aft
4. AIR REFUEL Switch	OPEN	Opens slipway door. Places FLCS in takeoff and landing gains when airspeed is below 400 knots
		Enables AR lights
		Reduces internal tank pressurization, depressurizes external tanks, and allows the refuel valve in each reservoir to open when a centerline tank is installed and refuel pressure is applied
	CLOSE	Reverses the OPEN actions

Sgt Howard Sacumo DCC SSgt Geronimo Gra

RESCUE

1. PUSH BUTTON TO OPEN DOOR
2. PULL RING OUT 6 FEET TO
   JETTISON CANOPY

Chapter Seven

Flying the Viper: the pilot's view

Major Mike 'Lobo' Canfield explains just how to use the multi-role capabilities of the F-16CM to their fullest potential:

'The F-16 Fighting Falcon, which I will refer to as the Viper, is a sleek and sexy aircraft. I've logged over 1,000 hours combined as a wingman, flight lead, instructor pilot, and evaluator pilot. I've also flown 275 combat hours in the skies over Afghanistan during Operation Enduring Freedom.'

OPPOSITE Reclined at 30 degrees, surrounded by a bubble canopy from the elbows up, and with no forward canopy framing to get in the way of spotting the enemy, the Viper pilot enjoys arguably one of the most comfortable and best-designed cockpits of any fighter in the world. *(USAF)*

111

LEFT The F-16C Block 50/52 cockpit is well designed with HUD, two MFDs and flight controllers positioned naturally either side of the canted ejection seat. *(Hellenic Air Force)*

CENTRE Three 192nd FS, Virginia Air National Guard Unit, pilots step to their Block 30C Vipers at Sandston, Virginia. The squadron would eventually lose its F-16s as it integrated with the 1st Fighter Wing (transitioning from F-16s to F-22A Raptors). *(USAF)*

The Viper is a fighter pilot's aircraft, designed around the pilot, for the pilot. In the following paragraphs I'll give you a glimpse of what it's like to be inside the cockpit of the Viper.

The F-16 is a multi-role aircraft, a jack of all trades. We go into an enemy's backyard, wipe out their Air Force, suppress and destroy their air defences, kill their troops, blow up their vehicles, bomb their bases with precision or non-precision weapons, and fight our way back out, picking off any leftover aircraft on the way home. Doing this well requires countless hours of study, ground simulators and training flights … the training never stops. A fighter pilot's goal is to be the best, and in the Viper you need to be the best at many different things.

Mission planning

The complexity of the training mission dictates how much time is needed for mission planning. If it's a basic fighter manoeuvring (BFM) – dogfighting – then there is very little planning for the sortie. If it's a large forces employment (LFE) (consisting of 69+ aircraft of different types, opposing forces, ground forces and support assets), the planning

LEFT Taken in the mid-1980s, this photograph shows Viper pilots mission-planning in the days before computers made that process so much quicker and more effective. Today, software tools allow the pilots to automatically 'weaponeer' the target (decide which weapons, release profiles and fuse setting will have the best effects), create flight plans, automatically check for hazards and areas of high bird numbers, as well as a range of other things. *(USAF)*

could be in the works for months, with the details of individual flights worked out 24–48 hours in advance of the actual sortie.

Generally, the members of the flight meet the day prior to discuss a game plan for the following day. For a four-ship, the Flight Lead (known as 'One') will assign tasks to the members of the flight according to their position in the flight. The Flight Lead and Element Lead, or Deputy Flight Lead (Three), will discuss tactics and overall mission flow.

The wingmen ('Two' and 'Four') prepare the line-up cards (small printouts containing mission-critical information such as call signs, radio frequencies and code words, transponder codes etc), maps, target photos, data transfer cartridge (DTC) that transfers pertinent mission data such as radio channels, steer points, Link-16 data link settings, weapon load-out and fusing, airspace and route lines, and anything else that can be downloaded to the jet.

One is responsible for the flight and the tactics they will use, and will brief the flight just prior to the mission. Once again, this is a generalisation of mission planning since it all depends on the type of mission scheduled.

Brief

The brief starts two hours plus or minus 30 minutes prior to take-off (depending on the type of sortie). Prior to starting the brief, the Flight Lead (me, in this case) will have set up briefing boards, and the wingmen will have laid out all the mission materials for each respective member of the flight. All members should be in the briefing room and ready to start the brief five minutes before brief time. I close the door and give everyone a 'nuclear' time hack that synchronises the flight's watches, gets us all on the same page and sets the tone for the precision that will execute our mission. When the door closes, rank no longer applies. One is the boss. No one else speaks unless asked a question or given the opportunity to speak: One has everyone's undivided attention.

The pilot conducts a general overview of the mission, followed by 'motherhood'. Motherhood is the basic admin execution of the flight: weather, divert status, NOTAMs (notices to airmen – information about airspace

and airfield restrictions, typically), emergency of the day, ground ops, take-off, handling of emergencies, to/from the airspace, airspace description, landing, training rules (TRs) and special subjects. Most items can be briefed as 'standard' (meaning they are covered in detail in a written publication called 'Wing Standards'). If I have a different way of conducting a certain part of motherhood, this is the time to let my flight know.

TRs are safety rules dictated by regulations and written in blood. I will highlight and brief the TRs that are applicable to the current mission. All members are required to know every TR. Special Subjects are items that the leadership requires us to discuss prior to flight. 'Gs' are generally a Special Subject in the F-16 community because of the Viper's 9G capability. That will kill an unprepared or complacent pilot.

Tactical admin is covered next. This covers mission-specific admin items such as missile settings, bomb fusing, flight set-up, expected threat, enemy tactics, etc. In today's mission, we are on the second day of the war. We've been tasked to destroy an enemy's command and control (C2) buildings and petroleum oil lubrication (POL) sites. Known surface-to-air missile (SAM) sites have been destroyed, but there is a probability of a few unknown sites still existing in and around our target area. There is an airborne defensive combat air patrol (CAP) in the area and a few airfields on alert consisting of enemy F-16s and F-5 Tigers. We will self-escort to the target and back. A KC-135 tanker will air-refuel us prior to push and an E-3 AWACS will provide radar coverage. We will be loaded out with AIM-120 AMRAAMs, AIM-9X Sidewinders, and 20mm for air-to-air (A/A) engagements. Our air-to-ground (A/G) load-out will be 2 x GBU-12 PAVEWAY IIs for One and Two and 2 x GBU-38 JDAMs for Three and Four.

For the main brief, I cover the tactics and flight member contracts (ie who will be doing what) that we will be utilising in both A/A phases and A/G phases of the flight. I cover radar search responsibilities, visual lookout, A/A targeting and shooting contracts, and defensive reactions. I'll explain in more detail our route of flight, timing and attack window. Looking at the target photos, I start big-to-small, looking for significant landmarks or funnelling features to get our eyes on the target.

A zoomed-out photo shows that our target area is a remote base in the desert, lying in a valley between hills and a dry riverbed. Looking at zoomed-in pictures, I see that my target is an L-shaped building on the southern side of an intersection that lies just to the west of the dried riverbed. I go over the photos with each respective member of the flight to ensure we can all identify our targets, and then ask for questions.

Overall, brief time varies with the complexity of the mission and experience of the pilots in the flight. After the brief everyone should be on the same page, and all questions should be answered before strapping on the jet. It's best to resolve uncertainties on the ground at zero knots/zero g, rather than a have a theoretical question mark appear in the HUD while flying at 510 knots pulling 6 to 9g.

BELOW The walk-around checklist is extensive, but Viper pilots will eventually memorise it so that no guide is needed. The inspection demands that the pilot examines specific items of interest in addition to generally being observant for bangs, dents or leaks. *(USAF)*

Step

Roughly ten minutes prior to stepping to the aircraft, we get 'dressed' in the aircrew flight equipment (AFE) shop. The AFE shop is a locker room, with each pilot having his own locker for storing his flight gear. The sounds of your brothers and sisters bantering about the upcoming missions, weather, and family life can be heard alongside g-suit zippers zipping and the clinking metal of harness clips as they are donned.

Helmets are checked for functionality and put in the helmet bag. Checklists are accounted for. I ensure I have all the mission materials including DTC and DVR (to record the mission). Finally, gloves and piddle packs are placed in the g-suit pockets.

The flight meets at the operations desk (Ops Desk) for aircraft assignments and is briefed with any updates to airfield status, weather or diverts. This is an opportunity to check the assigned aircraft in the 'bird book' for any gremlins (avionic or mechanical issues that are a nuisance but not unsafe for flight) living in that particular jet, and possible techniques for how to alleviate the problem. Pilots enter notes in the 'bird book' from previous flights like, 'The radar doesn't work the first two times you turn it on, but after you turn it off for 30 seconds and turn it back on the third time it will work fine.' After finishing up at the Ops Desk, depending on how far the jets are parked away from the squadron, we either walk or ride in the pilot van out to the jet.

Ground ops

Approaching the jet is always a humbling experience for me. It is such a beautiful piece of weaponry and I'm honoured to have the privilege to fight in it. The crew chief of the jet is standing in front of the Viper as I approach. These guys and their squadron mates put their blood, sweat and tears into ensuring the aircraft is ready for me to fly. Without their dedication, technical prowess and professional pride I'd be nothing but a guy wearing a flight suit.

The crew chief snaps to attention and renders a salute, which I return. We shake hands and he hands me the aircraft forms. I scan over the

Exterior Inspection (Typical)

NOTE: Check aircraft for loose doors and fasteners, cracks, dents, leaks, and other discrepancies.

NOSE - A

1. FORWARD FUSELAGE:
    A. EXTERNAL CANOPY JETTISON D-HANDLES (2) - ACCESS DOORS CLOSED.
    B. PITOT-STATIC PROBES (2) - COVERS REMOVED.
    C. AOA PROBES (2) - COVERS REMOVED; SLOTS CLEAR; FREEDOM OF MOVEMENT CHECKED; ALIGNMENT CHECKED (ROTATE PROBES FULLY TOWARD FRONT OF AIRCRAFT (CCW ON THE LEFT; CW ON THE RIGHT) AND VERIFY BOTTOM SLOTS SLIGHTLY AFT OF 6 O'CLOCK AND TOP SLOTS FORWARD); SET IN NEUTRAL POSITION (BOTTOM SLOT AT 4 O'CLOCK ON THE RIGHT SIDE AND 8 O'CLOCK ON THE LEFT SIDE).
    D. STATIC PORTS (2) - CONDITION.
    E. RADOME - SECURE.
    F. ENGINE INLET DUCT - CLEAR.
    G. PODS AND PYLONS - SECURE (PREFLIGHT IAW T.O. 1F-16CM-34-1-1).
    H. EPU FIRED INDICATOR - CHECK.
    I. ECS RAM INLET DUCTS - CLEAR.

CENTER FUSELAGE & RIGHT WING - B

1. RIGHT MLG:
    A. TIRE, WHEEL, AND STRUT - CONDITION.
    B. LG SAFETY PIN - INSTALLED.
    C. DRAG BRACE AND OVERCENTER LOCK, BOLTS, NUTS AND COTTER KEYS - CHECK SECURITY.
    D. UPLOCK ROLLER - CHECK.
    E. DOOR AND LINKAGE - SECURE.
2. RIGHT WING:
    A. HYDRAZINE - LEAK DETECTOR - CHECK.
    B. EPU NITROGEN BOTTLE - CHARGED (REFER TO FIGURE 2-5).
    C. EPU OIL LEVEL - CHECK.
    D. HYD SYS A QTY AND ACCUMULA-TOR - CHECK.
    E. GUN-RNDS COUNTER AND RNDS LIMIT - SET.
    F. SECURE VOICE PROCESSOR - CHECK.

G. EPU EXHAUST PORT - CONDITION.
H. LEF - CONDITION.
I. STORES AND PYLONS - SECURE (PREFLIGHT IAW T.O. 1F-16CM-34-1-1).
J. NAV AND FORM LIGHTS - CONDITION.
K. FLAPERON - CONDITION.

AFT FUSELAGE - C

1. TAIL:
    A. ADG-CHECK.
    B. CSD OIL LEVEL - CHECK.
    C. BRAKE/JFS ACCUMULATORS - CHARGED (3000 (+/- 100) PSI).
    D. HOOK - CONDITION AND PIN FREE TO MOVE.
    E. VENTRAL FINS, SPEEDBRAKES, HORIZONTAL TAILS, AND RUDDER - CONDITION.
    F. ENGINE EXHAUST AREA - CONDITION.
    G. NAV AND FORM LIGHTS - CONDITION.
    H. VERTICAL TAIL LIGHT - CONDITION.
    I. FLCS ACCUMULATORS - CHARGED (REFER TO FIGURE 2-6).
    J. JFS DOORS - CLOSED.

LEFT WING & CENTER FUSELAGE - D

1. LEFT WING:
    A. FLAPERON - CONDITION.
    B. NAV AND FORM LIGHTS - CONDITION.
    C. STORES AND PYLONS - SECURE (PREFLIGHT IAW T.O. 1F-16CM-34-1-1).
    D. LEF - CONDITION.
    E. FUEL VENT OUTLET - CLEAR.
    F. HYD SYS B QTY AND ACCUMULA-TOR - CHECK.
2. LEFT MLG:
    A. TIRE, WHEEL, AND STRUT - CONDITION.
    B. LG SAFETY PIN - INSTALLED.
    C. DRAG BRACE AND OVERCENTER LOCK, BOLTS, NUTS AND COTTER KEYS - CHECK SECURITY.
    D. UPLOCK ROLLER - CHECK.
    E. DOOR AND LINKAGE - SECURE.
    F. LG PIN CONTAINER - CHECK CONDITION.
3. FUSELAGE:
    A. GUN PORT - CONDITION.
    B. IFF - CHECK.
    C. AVTR - CHECK.
    D. DOOR 2317 ENGINE AND EMS GO-NO GO INDICATORS - CHECK.
4. UNDERSIDE:
    A. NLG TIRE, WHEEL, AND STRUT - CONDITION.
    B. NLG PIN - VERIFY REMOVED.
    C. NLG TORQUE ARMS - CONNECTED, PIN SECURE AND SAFETIED.
    D. NLG DOOR AND LINKAGE - SECURE.
    E. TAXI LIGHT - CONDITION.
    F. LG/HOOK EMERGENCY PNEUMATIC BOTTLE PRESSURE - WITHIN PLACARD LIMITS (REFER TO FIGURE 2-5). CM-FMPM-1-0123X40

forms to guarantee all checks are completed and signed off, look for any applicable open write-ups, and ensure my weapons load-out is correct. I then conduct my walk-around with my crew chief at arm's reach ready to answer questions I may have about the jet.

While conducting my walk around, I'm hypnotised by the ritual of checking probes, intake, panels, gauges, struts and flight controls; the familiar sights and sounds helps me get in synch for the upcoming flight. The walk-around is ingrained into the fighter pilot during F-16 initial training. At first a checklist and an Instructor Pilot will guide the student through the first few walk-arounds. Eventually, as time goes by, a flow or 'habit pattern' is developed. As a pilot's experience grows, the walk-around gets smoother and faster, and the checklist is readily available but rarely opened.

Habit pattern development is an essential part to being a single-seat fighter pilot. All aircraft have checklists, but most aircraft have co-pilots and engineers to help the pilot accomplish the steps. In a single-seat fighter there is no one there to help inside the cockpit. Initially the checklist is used line by line, but ultimately, after many hours of practice, the fighter pilot develops habit patterns using different techniques (including cockpit flow, acronyms, number of switches to be actuated, etc) for each checklist or normal phase of flight. I specify 'normal' phase of flight because when an emergency happens the pilot will reference the checklist after completing memorised critical action procedures (CAPs) and getting the aircraft under control.

After the walk-around is complete, I climb the ladder and stow my gear. I always put everything in the same spot so I can get to it without looking and always know where everything is. Viper pilots call this 'setting up their nest'. Helmet on the rail, DTC in the slot, helmet bag rolled up and stowed, checklist in my g-suit leg pocket, and mission materials strapped to my knee.

Easing into the cockpit, I strap the jet on to my back. This is a very important concept to learn when flying the Viper. The plane is being strapped on; you are in control of the aircraft, not the other way around. With my g-suit connected, seat kit buckled to the harness,

lap belt connected and adjusted, parachute clipped into harness, oxygen hooked up, joint helmet-mounted cueing system (JHMCS) cable connected, helmet on, intercom plugged in, chin-strap fastened, rudder pedals adjusted, and with strap-in completed I take a quick look over to make sure nothing was missed. The crew chief may follow up the ladder and assist in strapping in. A quick handshake, a thank you, and the crew chief climbs back down and removes the ladder. I double-check the start time on the line-up card and my watch and wait for the exact second to start the aircraft.

ABOVE The Viper's crew chief will already have conducted a meticulous walk-around inspection before the pilot arrives, but he will nonetheless accompany the pilot for this final thorough check in case any questions arise. *(USAF)*

BELOW Once the exterior inspection is complete, the pilot climbs the ladder, inspects the ACES II, then places his helmet and mask and any mission planning materials in the cockpit. The close confines of the cockpit mean that some cockpit preflight checks – in addition to checking switch positions for the first time – are easier done while perched on the ladder. *(Steve Davies/FJPhotography.com)*

Flight Instruments (Typical)

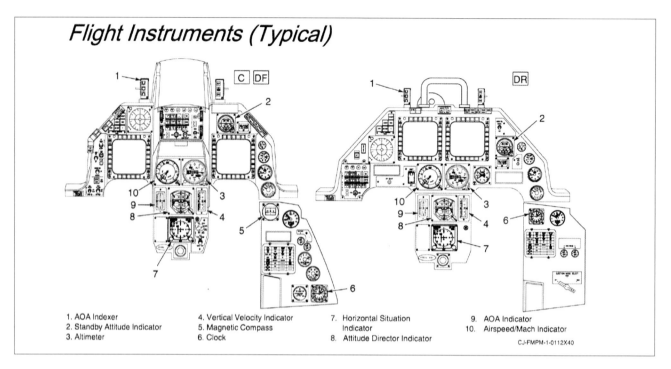

1. AOA Indexer
2. Standby Attitude Indicator
3. Altimeter
4. Vertical Velocity Indicator
5. Magnetic Compass
6. Clock
7. Horizontal Situation Indicator
8. Attitude Director Indicator
9. AOA Indicator
10. Airspeed/Mach Indicator

CJ-FMPM-1-0112X40

THIS PAGE AND OPPOSITE The F-16's instrument panels and control panels bunch controls and instruments into logical groups. Primary flight instrumentation is installed on the centre instrument panel, so that in the event of a HUD failure or when otherwise working heads down the pilot can still view flight attitude, airspeed, navigation and altitude information. The left auxiliary console provides landing gear, countermeasures and emergency jettison controls; the right features the caution light panel, pilot fault display list and oxygen, EPU, fuel and hydraulic dials. The left console is used to access radio, FLCC, lighting and electrical settings. The right console provides control over the Viper's sensors and oxygen systems. *(USAF)*

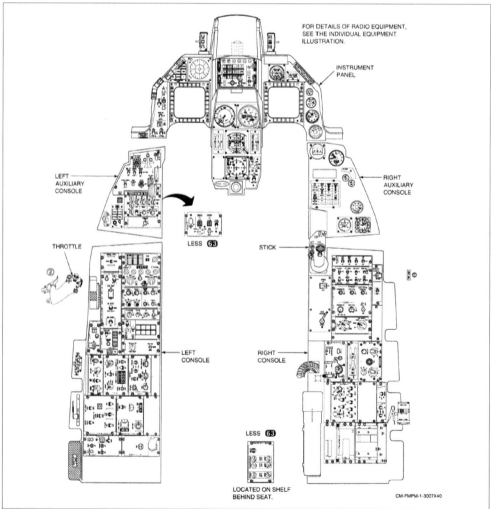

FOR DETAILS OF RADIO EQUIPMENT, SEE THE INDIVIDUAL EQUIPMENT ILLUSTRATION.

INSTRUMENT PANEL

LEFT AUXILIARY CONSOLE

RIGHT AUXILIARY CONSOLE

THROTTLE

LESS 63

STICK

LEFT CONSOLE

RIGHT CONSOLE

LESS 63

LOCATED ON SHELF BEHIND SEAT.

CM-FMPM-1-3007X40

INSTRUMENT PANEL

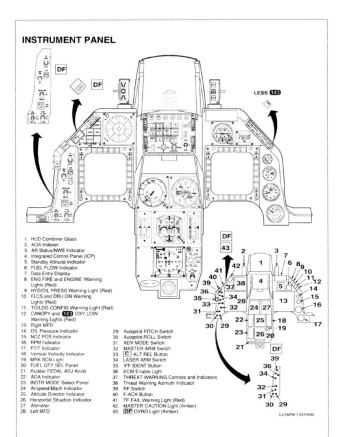

1. HUD Combiner Glass
2. AOA Indexer
3. AR Status/NWS Indicator
4. Integrated Control Panel (ICP)
5. Standby Attitude Indicator
6. FUEL FLOW Indicator
7. Data Entry Display
8. ENG FIRE and ENGINE Warning Lights (Red)
9. HYD/OIL PRESS Warning Light (Red)
10. FLCS and DBU ON Warning Lights (Red)
11. TO/LDG CONFIG Warning Light (Red)
12. CANOPY and 143 OXY LOW Warning Lights (Red)
13. Right MFD
14. OIL Pressure Indicator
15. NOZ POS Indicator
16. RPM Indicator
17. FTIT Indicator
18. Vertical Velocity Indicator
19. MRK BCN Light
20. FUEL QTY SEL Panel
21. Rudder PEDAL ADJ Knob
22. AOA Indicator
23. INSTR MODE Select Panel
24. Airspeed/Mach Indicator
25. Attitude Director Indicator
26. Horizontal Situation Indicator
27. Altimeter
28. Left MFD
29. Autopilot PITCH Switch
30. Autopilot ROLL Switch
31. ADV MODE Switch
32. MASTER ARM Switch
33. C ALT REL Button
34. LASER ARM Switch
35. IFF IDENT Button
36. ECM Enable Light
37. THREAT WARNING Controls and Indicators
38. Threat Warning Azimuth Indicator
39. RF Switch
40. F-ACK Button
41. TF FAIL Warning Light (Red)
42. MASTER CAUTION Light (Amber)
43. DF OVRD Light (Amber)

CJ-FMPM-1-0010X40

LEFT AUXILIARY CONSOLE | RIGHT AUXILIARY CONSOLE

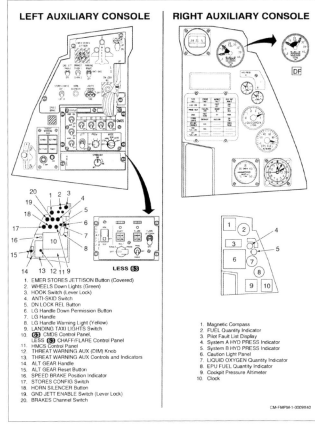

1. EMER STORES JETTISON Button (Covered)
2. WHEELS Down Lights (Green)
3. HOOK Switch (Lever Lock)
4. ANTI-SKID Switch
5. DN LOCK REL Button
6. LG Handle Down Permission Button
7. LG Handle
8. LG Handle Warning Light (Yellow)
9. LANDING TAXI LIGHTS Switch
10. 63 CMDS Control Panel, LESS 63 CHAFF/FLARE Control Panel
11. HMCS Control Panel
12. THREAT WARNING AUX (DIM) Knob
13. THREAT WARNING AUX Controls and Indicators
14. ALT GEAR Handle
15. ALT GEAR Reset Button
16. SPEED BRAKE Position Indicator
17. STORES CONFIG Switch
18. HORN SILENCER Button
19. GND JETT ENABLE Switch (Lever Lock)
20. BRAKES Channel Switch

1. Magnetic Compass
2. FUEL Quantity Indicator
3. Pilot Fault List Display
4. System A HYD PRESS Indicator
5. System B HYD PRESS Indicator
6. Caution Light Panel
7. LIQUID OXYGEN Quantity Indicator
8. EPU FUEL Quantity Indicator
9. Cockpit Pressure Altimeter
10. Clock

CM-FMPM-1-0009X40

LEFT CONSOLE

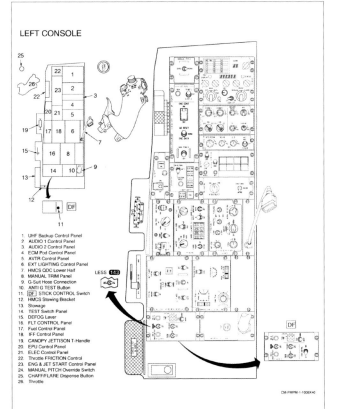

1. UHF Backup Control Panel
2. AUDIO 1 Control Panel
3. AUDIO 2 Control Panel
4. ECM Pod Control Panel
5. AVTR Control Panel
6. EXT LIGHTING Control Panel
7. HMCS QDC Lower Half
8. MANUAL TRIM Panel
9. G-Suit Hose Connection
10. ANTI G TEST Button
11. DF STICK CONTROL Switch
12. HMCS Stowing Bracket
13. Stowage
14. TEST Switch Panel
15. DEFOG Lever
16. FLT CONTROL Panel
17. Fuel Control Panel
18. IFF Control Panel
19. CANOPY JETTISON T-Handle
20. EPU Control Panel
21. ELEC Control Panel
22. Throttle FRICTION Control
23. ENG & JET START Control Panel
24. MANUAL PITCH Override Switch
25. CHAFF/FLARE Dispense Button
26. Throttle

CM-FMPM-1-1008X40

RIGHT CONSOLE

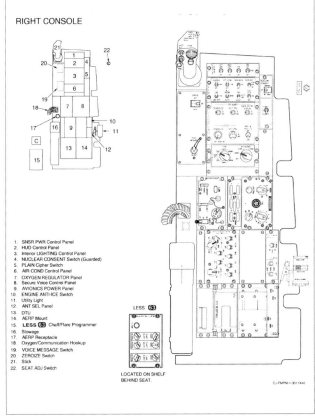

1. SNSR PWR Control Panel
2. HUD Control Panel
3. Interior LIGHTING Control Panel
4. NUCLEAR CONSENT Switch (Guarded)
5. PLAIN Cipher Switch
6. AIR COND Control Panel
7. OXYGEN REGULATOR Panel
8. Secure Voice Control Panel
9. AVIONICS POWER Panel
10. ENGINE ANTI-ICE Switch
11. Utility Light
12. ANT SEL Panel
13. DTU
14. LESS 63 Chaff/Flare Programmer
15. Stowage
16. AERP Receptacle
17. Oxygen/Communication Hookup
18. VOICE MESSAGE Switch
19. ZEROIZE Switch
20. Stick
21. SEAT ADJ Switch

LOCATED ON SHELF BEHIND SEAT.

CJ-FMPM-1-3011X40

Exterior Lighting (Typical)

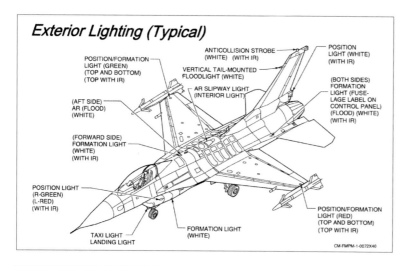

ANTICOLLISION STROBE (WHITE) (WITH IR)

POSITION/FORMATION LIGHT (GREEN) (TOP AND BOTTOM) (TOP WITH IR)

VERTICAL TAIL-MOUNTED FLOODLIGHT (WHITE)

POSITION LIGHT (WHITE) (WITH IR)

(BOTH SIDES) FORMATION LIGHT (FUSE-LAGE LABEL ON CONTROL PANEL) (FLOOD) (WHITE) (WITH IR)

AR SLIPWAY LIGHT (INTERIOR LIGHT)

(AFT SIDE) AR (FLOOD) (WHITE)

(FORWARD SIDE) FORMATION LIGHT (WHITE) (WITH IR)

POSITION LIGHT (R-GREEN) (L-RED) (WITH IR)

POSITION/FORMATION LIGHT (RED) (TOP AND BOTTOM) (TOP WITH IR)

FORMATION LIGHT (WHITE)

TAXI LIGHT LANDING LIGHT

CM-FMPM-1-0072X40

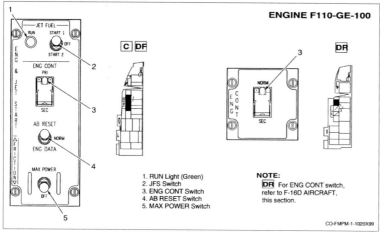

ENGINE F110-GE-100

1. RUN Light (Green)
2. JFS Switch
3. ENG CONT Switch
4. AB RESET Switch
5. MAX POWER Switch

NOTE:
DR For ENG CONT switch, refer to F-16D AIRCRAFT, this section.

CO-FMPM-1-1020X99

TOP Exterior light settings provide not only a visual indication that the F-16 is being cranked, but also indicate to other members of the flight whether the pilot is ready to taxi and/or take off. *(USAF)*

ABOVE The F110 start-up panel is nearly identical to that of the F100. Note the two JFS start modes and the green JFS run light. *(USAF)*

I power up the battery, check that the appropriate lights are lit, and move the power switch to 'MAIN POWER'. I check that I have good intercom communication with the crew chief on the ground, make sure it's clear to start the jet and lower my canopy. I move the jet fuel starter switch to 'START 2' and the engine begins to spool up. I push the throttle to 'IDLE' (or 'over the hump'), feel a slight thump as the fuel ignites and the engine lights, and monitor the engine gauges as the engine's rpm and FTIT (fan turbine inlet temperature) rise. With my hand resting on the throttle, I am ready to shut down if a problem arises during start.

As the engine purrs to life I can feel the controls snapping into position as hydraulic pressure rises, the JFS automatically shuts down, and the standby and main generators come online. I adjust my seat height and check the warning, caution and other lights on various panels throughout the cockpit and ensure engine instruments are reading normal. During

the remaining ground operations, the crew chief ducks and weaves around the jet, avoiding moving 3,000psi hydraulic-actuated flight controls. While checking gauges, flight controls and other checks, my fingers glide across buttons and switches in a habitual dance learned over thousands of repetitions. After five to ten minutes of checks, all my systems are on and operating satisfactorily.

I set up my avionics for Navigation (NAV), A/A, and A/G modes. Every pilot has his own technique to set up his avionics. These modes are switched through using the three-position Dogfight/Missile Override switch.

For the outboard position (or Dogfight mode), I put my AIM-9X as the main missile and gun set-up, with the Radar (RDR) and SMS (stores management system) in my left MFD and my horizontal situational display (HSD) and TST (Test) in the right MFD. I like to have the AIM-120 called up on the SMS in Missile Override (inboard position) with both MFDs displaying the same as in Dogfight.

The centre position defaults to the current Master Mode selectable on the upfront control panel (UFC) whether NAV, A/G or A/A. I have my A/A master mode set up with AIM-120 and the target (TGP) screen in the right MFD for long-range identification of aircraft. The A/G mode is set up with the HSD and SMS page on the left MFD and the TGP and RDR in the right MFD. Having the HSD in the opposite MFD helps me focus on dropping bombs.

I generally set up my NAV mode with RDR, DTE (data transfer equipment) and TST in my left MFD and HSD, TGP and SMS in the right. The pages are available in each MFD by flicking the display management switch left or right (depending on which MFD you want to change). I know it may sound confusing, but after setting up these modes the same way hundreds of times, it's as simple to navigate these modes as turning on your turn signal or shifting gears in your car.

I check the flight in to the second of the briefed check in time, 'Viper One, check', and hear the dutiful cadence 'Two', 'Three', 'Four'. I request taxi with a call to the ground controller. Once cleared for taxi, I signal my crew chief. He marshals me out of my spot, waves me off, and snaps another sharp salute followed closely with a friendly

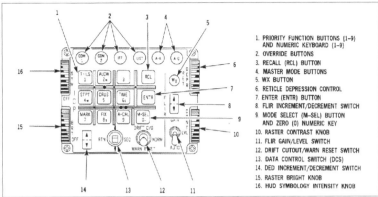

1. PRIORITY FUNCTION BUTTONS (1–9)
   AND NUMERIC KEYBOARD (1–9)
2. OVERRIDE BUTTONS
3. RECALL (RCL) BUTTON
4. MASTER MODE BUTTONS
5. WX BUTTON
6. RETICLE DEPRESSION CONTROL
7. ENTER (ENTR) BUTTON
8. FLIR INCREMENT/DECREMENT SWITCH
9. MODE SELECT (M-SEL) BUTTON
   AND ZERO (0) NUMERIC KEY
10. RASTER CONTRAST KNOB
11. FLIR GAIN/LEVEL SWITCH
12. DRIFT CUTOUT/WARN RESET SWITCH
13. DATA CONTROL SWITCH (DCS)
14. DED INCREMENT/DECREMENT SWITCH
15. RASTER BRIGHT KNOB
16. HUD SYMBOLOGY INTENSITY KNOB

squadron signal (a hand raised and curled in the shape of a claw, symbolising the squadron mascot … a vicious terrorist-killing buzzard).

The Viper is controlled using selectable nose-wheel steering (NWS), selected by pressing the 'MISSILE STEP' button on the stick, and when it's activated I steer using the rudder pedals. I won't use any more than 80% rpm to get the jet moving, and I always ensure my throttle is at idle and the nozzle has swung open prior to turning, to avoid my jet blast knocking over any vehicles or people.

ABOVE AND LEFT The UFC allows the pilot to key in data before taxi and take-off. Note the digital entry display – the small screen with green text to the right of the UFC. The pilot's inputs are visible on this display, although the information it shows can also be repeated in the lower section of the HUD. Note also the colour MFDs (radar displayed on the left MFD, horizontal situation display on the right) that are designed to be read even in bright sunlight. *(Hellenic Air Force/USAF)*

BELOW There are many hazards for ground crew working around the jet once the engine is cranked and the aircraft comes alive. These two diagrams show the danger areas for Block 42 and 52 Vipers. With maintainers scurrying around the aircraft during the EOR checks, the pilot will ensure that his hands and feet are clear of the controls to avoid any flight control surface movements. *(USAF)*

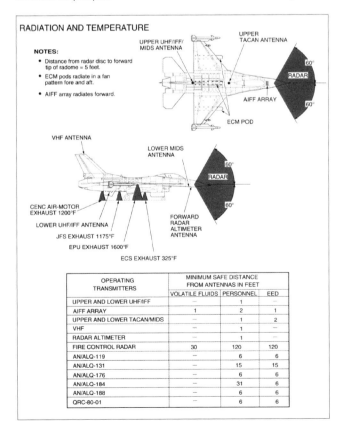

RADIATION AND TEMPERATURE

NOTES:
- Distance from radar disc to forward tip of radome = 5 feet.
- ECM pods radiate in a fan pattern fore and aft.
- AIFF array radiates forward.

OPERATING TRANSMITTERS	MINIMUM SAFE DISTANCE FROM ANTENNAS IN FEET		
	VOLATILE FLUIDS	PERSONNEL	EED
UPPER AND LOWER UHF/IFF	–	1	–
AIFF ARRAY	1	2	1
UPPER AND LOWER TACAN/MIDS	–	1	2
VHF	–	1	–
RADAR ALTIMETER	–	1	–
FIRE CONTROL RADAR	30	120	120
AN/ALQ-119	–	6	6
AN/ALQ-131	–	15	15
AN/ALQ-176	–	6	6
AN/ALQ-184	–	31	6
AN/ALQ-188	–	6	6
QRC-80-01	–	6	6

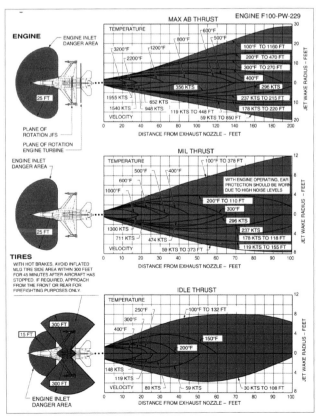

The four-ship taxis out in accordance with the flight brief, keeping precise spacing. We pull into the hammerhead for our end-of-runway (EOR) checks. A crew chief marshals us into position, and the EOR crew – consisting of weapons troops and crew chiefs – conduct final checks and weapons arming. My hands remain clear and visible to the crew chief standing out front, to ensure I don't bump any flight controls and inadvertently hurt one of the troops under my wing. Once this is complete we taxi to the runway and await our take-off time.

Take-off

Finally, it's time to take the Viper airborne. There are a lot of airspeed numbers associated with the take-off and landing data (TOLD) in any aircraft, and the Viper is no different. Additionally, there are different configurations I may be flying from day to day: clean jet (no external fuel tanks or bombs), single-bag jet (external fuel tank on the centreline), two-bag jet (two external fuel tanks on wing stations 4 and 6), etc. The type of take-off, whether in MIL power or afterburner (AB) mode, is dependent on the aircraft configuration, runway length, pressure altitude and temperature.

I line up the jet on my side of the runway and my wingman does the same on his side, with Three and Four lined up behind. Quick hand signals and acknowledgements flow up and down the formation in sequence for the engine run-up. We push our throttles up to 80%. After

I note that no one is creeping due to poor brakes, I glance down the engine instrument stack to ensure everything is good to go. I look at my wingman, who by now is looking at Three and waiting for the 'big nod' that he is good to go (which he will communicate after receiving a 'big nod' from Four).

My wingman receives the nod and repeats it to me. I salute him off (in this case we will be doing a four-ship interval take-off), release brakes and push the throttle into AB. Another quick glance at my engine stack ensures a good nozzle swing as my AB lights and I feel the kick of acceleration pushing me back against the seat. Hurtling down the runway, the airspeed comes alive in the HUD (around 60kt), I click off the nose-wheel steering and use rudder to maintain directional control, though not much is ever needed.

I can hear the mechanical clicking of the fuel flow indicator as it tries to catch up with the actual amount of fuel being thrown into the fire (50,000+lb an hour). I throw another glance at the engine stack, 'eyebrow' warning panel and caution panel as I approach refusal speed (the speed at which I can successfully abort on the remaining length of runway), for any signs not to take the jet airborne. With everything looking good, I apply a small amount of pressure to the stick at rotation speed and pull the bore cross to 10° nose high. 15kt later the aircraft is airborne. From brake release until now, roughly half a mile of runway has passed beneath me and 12 seconds have elapsed.

I retract my gear and continue my climb. Considering the noise the F-16 makes during an AB take-off, the cockpit is relatively quiet. My vertical velocity indicator (VVI) – though in honesty I rarely look at this instrument – is pegged at 6,000ft a minute. I slowly pull the throttle out of AB at 300kt, continue to accelerate to 350kt and double-check my engine instruments. I honestly don't know how often I look at my engine instruments during a flight, but I assume it's often since my livelihood is dependent on that one engine!

I continue to climb to my assigned altitude and make any radio calls required. I glance over my shoulder to check on my wingman's rejoin. He released brakes 15 seconds after me, which is standard for an AB take-off in good weather,

BELOW Afterburner take-off in the F-16C/D is a fast-paced and dizzying affair: a lot happens very quickly, and the pilot must make sure that the engine is within limits and operating as advertised. Meanwhile, the raw power in the base of the spine provides longitudinal acceleration forces that never fail to impress. This Nellis-based Aggressor F-16 leaves the runway with afterburner raging and heat plume trailing. (USAF)

with Three and Four following suit. If there is bad weather we will go to radar-assisted trail departure. In that case, Two will wait 20 seconds for brake release, and after climbing safely from the ground and retracting his gear he will use his radar to lock me up and keep a safe distance until we are above the weather and able to join up visually. We also have the option to do a two-ship formation take-off, with Three and Four following 15 seconds behind.

As the Flight Lead, I clear for the flight, keeping a scan going for potential conflicts popping up on the radar scope in the upcoming airspace (known as 'far rocks'), and also looking outside the jet for immediate threats ('near rocks'), birds or airborne traffic conflicts.

I will keep a constant crosscheck going throughout the entire flight, never pausing, prioritising different items depending on the phase of flight. We have a helmet-mounted camera that we can record and watch during debrief, but, as much as we move our craniums around during flight conducting our crosscheck, it is very dizzying to watch, and could cause motion sickness to the viewer.

Earlier I mentioned the Viper was designed for the pilot. While this may not be 100% true, sitting in the Viper during flight sure makes me feel like they had me in mind while designing

her. The seat is reclined back at a comfortable 30°; the throttle is nestled comfortably in my left hand; my right hand is on the stick, with an armrest supporting my right arm. It's almost as comfortable as sitting in a leather recliner, beer in one hand, TV remote in the other.

The bubble canopy is the best 'office with a view' anyone could ask for. It provides unbelievable visibility and breathtaking views. The bottom of the canopy extends down to just above my waist. It then bubbles out prior to terminating into the canopy rail, allowing me to look out and underneath my jet. The cockpit sits far forward on the Viper. There is no intake

ABOVE A formation of Vipers from the 421st FS, Hill Air Force Base, Utah, climbs to cruise altitude en route to the target. The flight lead, nearest the camera, is responsible for navigation, while the entire flight is responsible for checking for threats ahead of and behind them. (USAF)

BELOW Even the view from the back of the Viper is impressive. Designed originally as a lightweight fighter, visibility is crucial once air combat enters the visual arena. The saying 'lose sight, lose the fight', is written in blood. (Hellenic Air Force)

BELOW A two-ship of 510th FS Block 40s ascends through cloud into the sunshine. With the stick moving only a very small amount, formation flying the Viper is less about actually moving the stick and more about applying tiny amounts of pressure to it. (Steve Davies/FJPhotography.com)

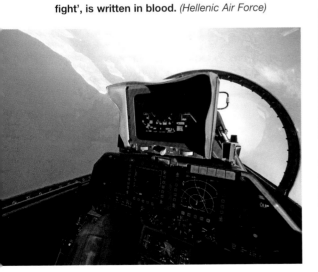

nor wings preventing me from seeing almost directly below.

Unlike many aircraft, there is also no canopy bow to obstruct my view. The fly-by-wire side-mounted stick is brilliant. The stick only moves a quarter of an inch and activates the control surfaces by monitoring the amount of force I'm inputting, unlike conventional cable-and-pulley controls which respond to the amount I'm moving the stick. I check my engines and fuel and prompt my wingmen to do the same over the interflight radio frequency. This verifies our systems are operating normally while we are still close to base. I'll do more fuel checks throughout the sortie at regular intervals to guarantee we don't overfly our 'bingo' fuel (ie the point at which only enough fuel remains to safely return to base without using reserve fuel).

Two is finishing his rejoin to route formation by now. I give him the signal to assume tactical spread formation (one-mile line abreast) by 'porpoising' my aircraft (rapid up and down movements like a dolphin swimming). Whenever possible, I want my wingman to be far enough

away from me to be able to monitor his aircraft systems, work the radar, and visually scan the area for threats (birds and other aircraft in peacetime). Now Three calls 'Saddled', meaning he is at a pre-briefed formation – in this case one mile in trail of me, with Four in tactical spread off of Three.

En route

Levelling off, I push almost imperceptibly on the stick while retarding the throttle. I set a fuel flow to maintain 350kt as we cruise to our assigned airspace. The Viper is trimmed automatically to 1g level flight, meaning once I level off I theoretically don't have to do anything else with the flight controls to keep level. I *do* have to monitor airspeed, and give slight tweaks periodically to the stick and throttle to maintain level flight and constant airspeed. This is nothing near the workload of flying most aircraft with trimming and re-trimming. We do have an autopilot system with altitude and attitude-hold capabilities, but I rarely use it to and from the airspace unless utilising a piddle pack, getting out an instrument approach plate, or something else that requires my attention.

The Viper is relatively easy to fly; in fact our student pilots are solo in the Viper on their fourth sortie in one. The difficulty in piloting the Viper isn't stick and rudder – the engineers did a great job creating a stable, pilot-friendly aircraft. The difficulty arises when employing it successfully as a weapon.

The purest moments I have ever experienced have been flying the Viper. It requires all my concentration and attention. There are no stray thoughts of bills to pay, meetings to attend, project deadlines or other matters that wander into my mind. My entire being is absorbed into accurately employing my four-ship.

When I first started out flying in the Air Force in the now retired T-37 Tweet, I would often become task-saturated with just one or two tasks, like keeping the aircraft level or maintaining the correct airspeed. Guided by great instructors and a ton of practice, I was able to expand my crosscheck to six or nine things.

Eventually I was trusted into taking the Tweet airborne solo. This crosscheck progression continued into the T-38 Talon and eventually

BELOW With the ACES II seat canted back 30°, the Viper's seating position is particularly comfortable when under high g loads. This diagram shows how the canopy sill line runs well below the shoulder, increasing overall visibility from the cockpit. *(USAF)*

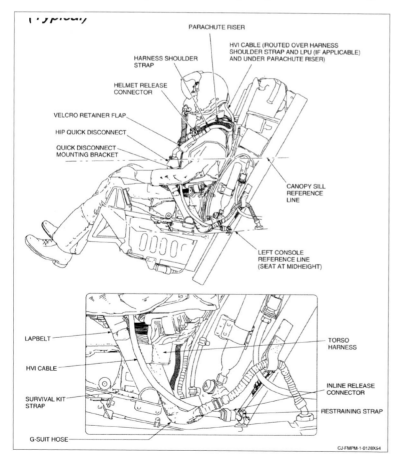

PARACHUTE RISER

HVI CABLE (ROUTED OVER HARNESS SHOULDER STRAP AND LPU (IF APPLICABLE) AND UNDER PARACHUTE RISER)

HARNESS SHOULDER STRAP

HELMET RELEASE CONNECTOR

VELCRO RETAINER FLAP

HIP QUICK DISCONNECT

QUICK DISCONNECT MOUNTING BRACKET

CANOPY SILL REFERENCE LINE

LEFT CONSOLE REFERENCE LINE (SEAT AT MIDHEIGHT)

LAPBELT

HVI CABLE

SURVIVAL KIT STRAP

G-SUIT HOSE

TORSO HARNESS

INLINE RELEASE CONNECTOR

RESTRAINING STRAP

CJ-FMPM-1-0128X54

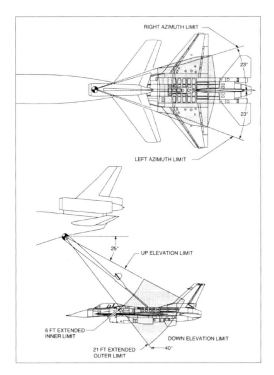

The F-16 can refuel from the boom of either the KC-10 or the KC-135. In these three illustrations, the KC-10 is illustrated. A shaded area on the boom limits drawing [above] shows the vertical and horizontal limits of the boom: 23° in azimuth either side of centreline, and from 25° down to 40° down from the centreline. Note the director lights in the boom markings [above right] and KC-10 markings [right] images – these guide the pilot into the optimum position for refuelling. *(USAF)*

the Viper. Each new aircraft introduced new tasks to perform during the crosscheck, or a different way to complete them.

Hundreds of hours of studying and flying, along with patient, demanding instructors who taught me the procedures and their individual techniques, eventually resulted in my efficiency in the jet and my competency to lead other aircraft into combat.

Refuelling

Levelling off at 24,000ft with clearance into the airspace, I visually check my wingmen's positions, push the flight over to the tanker frequency, and contact the KC-135 that is established in an orbit waiting for our rendezvous.

My flight's positions are displayed on the HSD

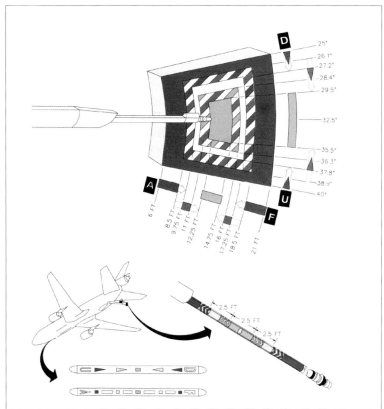

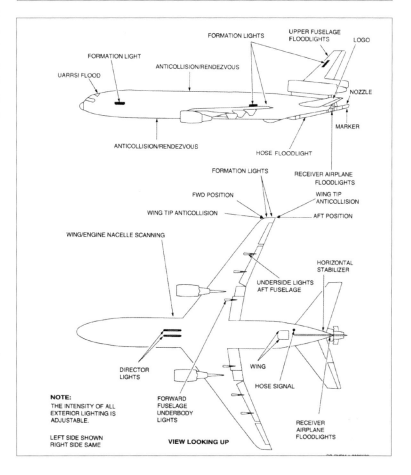

ABOVE Air refuelling
is a critical component
for almost any mission.
The F-16 is small and
has a correspondingly
small fuel capacity,
meaning that 'tanking'
successfully will
make the difference
between going on to
the target or returning
home. Here, two
F-16 Block 50s from
Spangdahlem's 22nd
and 23rd FSs take on
fuel before heading
into Iraq. (USAF)

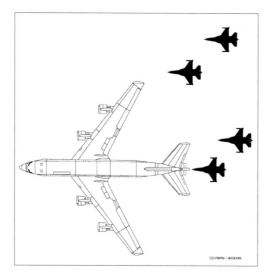

CO-FMPM-1-8008X89

RIGHT Flowing to and
from the tanker quickly
is also essential to
ensuring that everyone
gets fuel and can push
into enemy territory
together and on time.
Recipients will flow,
as this illustration
shows, from the right
wing to the left as they
sequence on and off
the boom. (USAF)

as well as on my JHMCS visor. In the helmet, a green circle is superimposed on them showing their range and number. After determining the tanker's position, I move the cursor control button with my thumb, place my radar cursors over the tanker's suspected location, press the TMS (target management switch) forward, or up, to lock him, note his altitude, airspeed, range and aspect, and set up my intercept.

In my HUD and helmet the radar lock puts a square over the tanker, helping with visual pick-up. I call for the flight to conduct their Pre Air-Refuelling checklists, and for Two and Four to rejoin on their element leaders. I make slight adjustments to my intercept angle to 'cool down' the intercept, making it easier for my wingman to stay in formation.

I pick up the tanker visually at around ten miles. It's a looming black dot on a brilliant blue sky, smack in the middle of the green radar-lock square. I request to join and continue to work my rendezvous to arrive 1,000ft behind the waiting boom. With the intercept near complete and everyone in a stable position, I tell my three wingmen to rejoin to observation position. Observation position is 15ft or so off the left wing of the tanker, and high enough to see the opposite wing's outboard engine, which keeps them out of the way so as not to interfere with the aircraft that is on the boom refuelling.

I stabilise into position 50ft behind the KC-135, matching airspeed (310kt) and request to 'contact'. The boomer, who is 'flying' the boom and watching me out of the aft window of the tanker, clears me into contact position. I ease my throttle forward slightly and accelerate slowly forward with the refuelling boom in the middle of my HUD.

The boom is roughly 30ft long and 10ft in diameter. The best approach to it is at a 'fast walk' pace. I don't stare at the boom, but my eyes never stray far from it. I make sure I have a light grip on the stick so that I don't over-control my jet – it is very sensitive in pitch, and I would be unwise to make big inputs into the controls this close to a 300,000lb aircraft! I use my fingertips on the stick only, and am cognisant of any muscles tensing so that I can quickly force myself to relax.

As I approach the boom, the jet wash from the tanker's engines can be felt on my wings, and it pushes the aircraft down slightly. I adjust up somewhat and continue forward, getting past the jet wash. As I move in closer I can actually hear the roar of the tanker's four turbofans. Just prior to the point where I think I'm going to hit the boom with my nose, the boomer moves it to the left. Every now and then a boomer moves it right, causing my eyelids to initially widen just a little at this change in habit. Though it really doesn't make a difference, I am curious if there is a reason for the switch-hitting.

The boom is now in my peripheral vision out of the side of the cockpit. I continue forward, slightly adjusting my momentum to a slow-walk speed. I never turn and look at the boom, as it would probably cause me to fly into it. Once the boom is clear of the canopy the boomer will move the boom in line with my fuselage and plug it into the fuel port in my jet.

I feel a slight jostle in the aircraft as the boom connects, and take a mental snapshot of where the tanker is in my field of view. I use this technique to maintain a steady, constant position. There are also director lights on the belly of the tanker that will give me arrows (forward, aft, up, down) to help me maintain my position. There are no left or right arrows since I should be able to keep our fuselages aligned.

I can now speak to the tanker directly with a 'hot mike' through a comm connection in the boom. I usually ask where they're from and how they're doing. Once the fuel is flowing in I will check my fuel gauge out the corner of my eye to ensure my fuel quantity is increasing. I'm keeping my body movements minimal to avoid any unintended inputs that would affect my position. I continue to adjust using pressure from my fingertips only, resisting the urge to move my throttle. Obviously, if throttle adjustments are required I make them small.

Once the required amount of fuel is transferred, I thank the boomer, retard my throttle and drift aft and low, clearing the boom, slide to the right and accelerate to join up on the tanker's right wing. I then monitor my wingmen as they refuel in sequence. Once Four is in contact with the boom, I'll coordinate with the tanker pilot on the radio for instructions to depart the tanker track and proceed to our assigned airspace.

Fight's on

Entering the airspace, I get the flight ready to fight by calling 'Viper Fence-in'. The Fence Check is a checklist of items that are memorised by the pilot to turn on the Viper's offensive and defensive weapon systems: A/A mode selected, missiles ready, chaff/flare armed, oxygen mask clicked up, radar warning receiver (RWR, pronounced 'raw') threat volume turned up, and all other tactical equipment on and ready.

I have the flight in tactical formation and perform the 'G-Check.' This is a 90° turn pulling 4–5g, and then a 180° turn pulling 6–7g (not with bombs, since they are g-limited to 5.5g). This check is to ensure that your g-suit and helmet inflate and your mask provides pressurised breathing to help get more oxygen in your body while under g.

This is also the chance to evaluate your own body's performance, and how your g-tolerance is for the day in a benign environment. You don't want to realise your body or jet isn't ready for gs only as you're merging with an enemy aircraft! When under g, the blood in your body is pulled to the lower extremities, starving your brain of blood and oxygen. Without preventative measures you'd lose consciousness – suffer GLOC (g-induced loss of consciousness) – with potentially deadly results.

Performing the anti-g strain manoeuvre (AGSM) is another learned skill developed early on in fighter training. A proper AGSM is performed by taking a deep breath, closing off your glottis and flexing all your lower muscles (feet, calves, thighs, butt and abs), forcing the blood back into your brain. As pressure builds in your chest, you relieve this pressure with a quick partial exhale/inhale. The g-suit helps the pilot by inflating and squeezing the legs and abdomen, assisting in pushing the blood back into the brain.

Pulling gs is fun – to a point. For me, that point is 7.5g; anything above 7.5g physically hurts and is exhausting. However, if pulling 9g gives me the advantage in a dogfight, I'd rather be able to shoot than get shot.

I proceed to our ingress point that was loaded from the DTC, and use the avionics to monitor our progress, confirming that we will make the planned 'push' time for our fight. All flight members are 'fangs-out' – ready to kill. My crosscheck continues as I monitor my radar, wingmen and jet systems. I check in with AWACS and let them know we are 'as fragged',

ABOVE Prior to actually pushing in to enemy airspace, a g-check will ensure that the g-suit is working and that the pilot is 'warmed up'. The Viper creates streamers and swirling vortexes as it pulls high gs. *(USAF)*

meaning we've arrived as planned. If I've had any changes, I let them know, and they in turn advise me of any changes to the attack plan that they may have received from headquarters.

They also let us know that they are in radar contact with us and give us any updates on pertinent friendly and enemy activity.

We reach our ingress point and I push my flight into tactical spread, or 'wall' formation. We climb and accelerate to stores airspeed limits, scanning our radars for threats. At this point I feel very offensive, leaning forward in my seat and begging for a fight.

AWACS calls out multiple enemy aircraft contacts in our path to the target area. My flight's radars are locked in, filling my helmet's visor with green squares. As they launch their AIM-120 AMRAAMs at the targets beyond visual range (BVR), I hear 'Fox 3' over the radio. I ascertain that all aircraft that are a factor have been targeted and shot at by using my avionic displays and listening to the radio calls from my flight. In a training scenario, nothing actually leaves the aircraft. However, we do have electronic systems and procedures that will help us determine a valid 'kill'. Our missiles time out (meaning they reach their targets based on range), and the initial wave of enemy fighters is destroyed.

We keep our radars scanning the area and with the help of AWACS make certain the airspace remains clear. While shooting BVR is not as exciting as old-fashioned dogfighting, it is nice to be able to clear our path at a distance without wasting time and fuel over enemy territory.

Employing the Viper's arsenal of weapons has been simplified with our hands-on throttle and

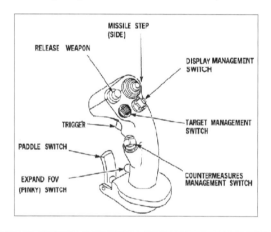

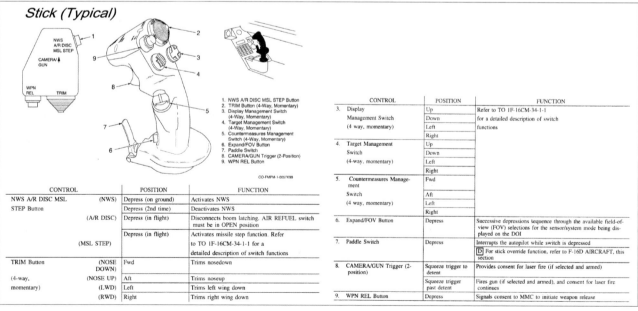

stick (HOTAS) design. The flight stick has nine switches and the throttle six. Learning to use the HOTAS takes practice at first, but it is intuitively designed. Now, my fingers move around the switches without much thought, like learning to type or, more relevant, using a game controller.

Proceeding on our planned route, we punch off our empty external fuel tanks (we don't punch the tanks off in training – it's too expensive) and accelerate. We are entering the enemy's air defence corridor. Most of the SAM sites have been destroyed in preceding missions, but a few are presumed to remain. My crosscheck will include my RWR more often now, as I scan my radar display and look outside of my cockpit for threats, from both the ground and air.

Two is flying good formation and I've cleared Three to take spacing. Three has manoeuvred himself and Four back to a pre-planned distance in trail as we approach our planned AGTR (air-to-ground transition range). This is a planned distance from the target at which we take off our A/A 'hats' and put on our A/G 'hats', in preparation to drop our bombs.

I take one last scan of the target area and beyond on my radar before I move my Dogfight/Missile Override switch to the centre position and press the A/G button on my UFC. As I mentioned in my section on ground ops, having the HSD and Radar on opposite displays helps me transition to a bomb-dropping mindset. I double-check my SMS page by flicking my display management switch left, and check that the correct bomb is selected with the correct fuse settings. I then flick the DMS switch again to bring the HSD back up.

I check that the right MFD displays the TGP and look to see if my target is discernible in the picture with the use of my pre-mission map study lead-in features. I locate the dry riverbed just short of a group of buildings at an intersection.

I zoom in on the intersection by spinning the hands-on gain (HOG) knob located on the throttle, and simultaneously thumb the cursor control to drag my pod on to the L-shaped building to the south. I click the pinky switch on the stick to change my camera lens in the pod to narrow field of view (NFOV), making my picture even more detailed. I zoom in again and refine my cursor placement to an airshaft on the roof. I

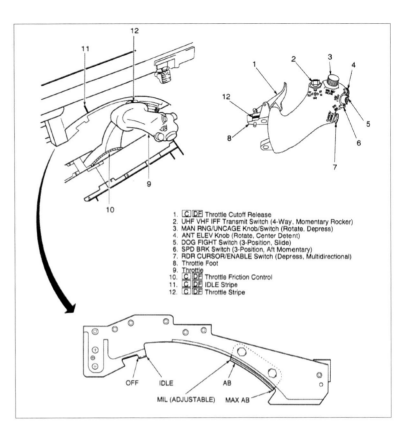

1. [C][DF] Throttle Cutoff Release
2. UHF VHF IFF Transmit Switch (4-Way, Momentary Rocker)
3. MAN RNG/UNCAGE Knob/Switch (Rotate, Depress)
4. ANT ELEV Knob (Rotate, Center Detent)
5. DOG FIGHT Switch (3-Position, Slide)
6. SPD BRK Switch (3-Position, Aft Momentary)
7. RDR CURSOR/ENABLE Switch (Depress, Multidirectional)
8. Throttle Foot
9. Throttle
10. [C][DF] Throttle Friction Control
11. [C][DF] IDLE Stripe
12. [C][DF] Throttle Stripe

ABOVE Most of the throttle's HOTAS switches control the radar, target pod, targeting, weapons selection and MFD cursors. *(USAF)*

BELOW Speed is life. Approaching the Mach and pulling gs before pitching up to toss a load of bombs at the target, strakes of ragged condensation flow off the wings of this F-16C. As soon as the bombs are gone, the CAT switch will be moved to I and the pilot will enter air-to-air mode in his mind and with his radar. *(Steve Davies/FJPhotography.com)*

ABOVE The Viper's small plan view makes it difficult to spot until it's often too late. Many an Eagle and Hornet pilot has fallen victim to an F-16 that has reached the merge unsighted. *(USAF)*

BELOW The antenna locations of the F-16CM. The AIFF array, also known colloquially as bird slicers, sits in front of the canopy and electronically identifies friendly aircraft squawking pre-set IFF codes. The radar warning receiver, whose antennae are placed far apart to allow triangulation on other radar emitters, provides the pilot with visual and audio signals of radiation sources in the airspace. *(USAF)*

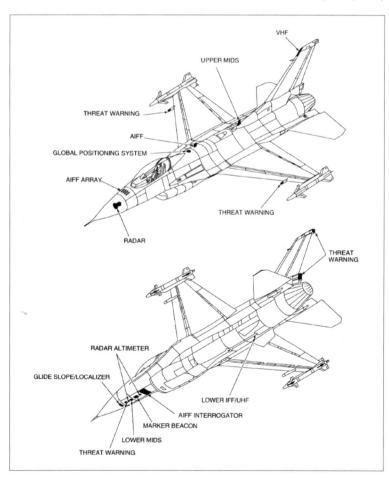

confirm my TGP screen matches the target photo located in my mission materials on my leg, and TMS right to put my pod in track mode.

I check my airspeed, altitude, timing and wingman's position and clear his six (the area immediately behind him) visually for threats. Everything is looking good, until I notice a bright flashing light in my cockpit and hear a loud warning in my ears. I focus my attention on the threat displayed on my RWR. While absorbing this information, ingrained instinct takes over. I throw my jet into defensive manoeuvres and press the CMS (countermeasures switch) to deploy countermeasures.

While pushing myself around in the jet, I twist in my seat, look outside for the missile launch, and communicate to my flight that I've been launched at by a SAM. The twisting and pushing is similar to turning your body when reversing your car. Considering I weigh more than five times my normal weight due to g forces, I have to put a hand on the towel racks' located on the canopy to help pull or push myself around. Bruises are common on my elbows after busy flights, from smacking the canopy or towel rack while twisting around in the cockpit.

Two flies to a supporting position and will point out the SAM if able, while Three and Four continue to their targets. I persist in manoeuvring until the missile is defeated or I'm dead. Thankfully, my RWR is no longer indicating a threat, and I haven't been called 'dead'. I reset course to flow in behind Three and Four, while Two flies back out to tactical position. I force myself to slow my breathing and steady my crosscheck.

Three and Four drop their weapons and begin their egress. My timing is a bit off now, and I double-check that we are still in our drop window. Accelerating back to my planned speed, I reacquire my target in the pod, double-check it with my target photo, TMS right to put the pod back in track mode, and prepare for release. Listening to the chatter over the radio it sounds like Three and Four are engaging enemy aircraft. I check that my steering is centred in the HUD and see the weapons release cue marching steadily down the steering line signalling the pending release. Approaching weapons release, I call out '3…2…1…weapon away.' I mash down on the 'pickle button' and feel the two GBU-12 PAVEWAY II 500lb bombs fall off my aircraft. The absence of the excess drag on my jet is immediately noticeable.

I check turn into my lase leg (the stage at which I must fire the target pod laser to guide my bombs) and glance over at my wingman, in time to see his own bombs falling away from his aircraft. He too manoeuvres his aircraft into his lase leg. I glance back inside my cockpit, confirm the target is still in my TGP display and fire the pod's laser to direct my bombs to the target. I look outside for any threats, check my RWR and look back at my pod display. Watching in the pod for the last few seconds, I see the satisfying plume as my bombs meet their target. I glance outside briefly to appreciate the destructive fireball and am rewarded with seeing Two's bombs impact his target a second later. The boy in me wants to enjoy the fireworks and stare until the fire plumes disappear. However, I also know we just kicked up a hornets' nest and need to get the hell out of there.

I thumb back into Missile Override mode, and with the radar back in the left MFD and HSD in the right, my A/A 'hat' is back on. I light the AB, feel the satisfying kick of thrust, begin a climb and move my Stores Configuration switch from CAT III to CAT I. This switch is an AoA and roll-limiting switch used to protect the Viper when loaded down with stores. In CAT I my jet is now an unlimited manoeuvring fighter (though g-limited by the pilot for the target pod).

I look off my wing and Two is spot-on in his welded tactical position. I see Three and Four 15 miles ahead of us on the radar and it sounds like they are on the offensive, engaging enemy aircraft visually. My scope looks clean other than that, and AWACS is not talking about any other threats.

'Viper One, break left! Bogey, your six o'clock, two miles, low.' I immediately execute a high-g break turn and wrench my body and head around to the left, trying to gain tally on the bogey that has jumped us. He rolls up his aircraft as he enters my turn circle. Bright orange missiles on his F-16 wingtips identify him as an enemy. 'Tally one, Hostile Viper.' I continue my spiralling dive towards the ground, watching the enemy aircraft, crosschecking

my airspeed and altitude in my helmet while trying to deny him a WEZ (weapon engagement zone).

The g forces pull sweat that has built up in my hair and eyebrows down into my eyes, and I try to blink the sting away. My wingman engages: 'Viper Two, FOX-2…kill Hostile Viper, left turn, bugout 090.' I continue my turn until I'm heading 090 towards home, push forward on the stick, causing me to float in my seat with the AB lit, and accelerate to get away from the fireball. Glancing over my right shoulder, I see as expected, Two flowing back into tactical position. You can always depend on a good wingman to be where you expect him to be. Reaching a good tactical airspeed, I climb to give us more potential energy to use if there is another fireball.

Checking my radar I locate a contact five miles ahead. I request an ID from AWACS, but unfortunately the contact has just recently popped up in the area. The required matrix to declare this contact an enemy has not been met. This means we have to go in and identify him visually (VID) before we can kill him. Three and Four are still fighting further ahead and are unable to assist at this point.

I target the enemy using my radar and set up the intercept, keeping an eye on my RWR and radar, watching for signs that he is aware of us. It looks like he is heading towards Three and Four. Transonic (about .98 Mach), I can feel shockwave push on my slabs as it travels further aft on my fuselage, making the jet a little more pitch-sensitive. It only lasts a second as I continue to accelerate supersonic. I look outside and see a black speck below us.

Rolling inverted (I never miss an opportunity

ABOVE Turning and burning with afterburner plugged in, a light Viper has an excellent thrust-to-weight ratio, allowing it to pull and sustain a 9g turn for as long as the pilot can handle it and as long as the fuel state allows. In combat, this capability can make the difference between life and death. *(USAF)*

to fly upside-down!), I pull my nose 20° low and roll back upright. 'Viper One, tally one, my nose, five miles low, black on brown,' I call into the radio. 'Tally one,' Two responds. I flick my DMS switch to the right and bring up my pod on the MFD to see if I can ID him. No such luck. At this distance, I want to be looking outside as much as possible, so I don't waste time trying to acquire him in my pod.

Scanning around the bogey I see a glint of sun flash off something, and see his wingman flying a mile off his wing. My wingman calls out 'Viper Two, tally-two, line abreast.' I respond with 'Viper One, tally-two.' They are still heading towards Three and Four, unaware that death is a few miles over their shoulders and closing fast. We arrive one mile off their eight o'clock; I see in my visor that Two has targeted the closer of the two aircraft. I switch into Dogfight mode, TMS forward. Holding the button forward, I bring the targeting oval into my helmet, thus slaving my radar to where I look. I place the oval over the far bogey and let go of the TMS button. Simultaneously I hear 'Bitching Betty' (an automated voice in the jet that passes confirmation messages) say 'Lock' and the targeting box appears over the bogey.

Preparing for a knife-fight in a phone booth, I slow to subsonic. If I were to stay supersonic, my turn radius would be huge and I would lose my jet's advantage in a dogfight. I fly over the top of the closer target I see it's an F-5, and radio 'Viper One, ID Hostile Tiger.' Two engages and kills the first aircraft with a missile shot as I engage the second enemy aircraft.

I've only launched one missile in my life. They aren't cheap, and we usually only launch them in combat or test sorties: I had the privilege of shooting an AIM-9LM Sidewinder at a drone. With the target locked up (a purr from the missile lets me know I have a good infrared lock), I mashed down the 'pickle button' and it seemed like nothing happened. Some pilots experience temporal distortion, and time appears to slow down. This was my experience as well. I kept the 'pickle button' pressed and looked over at the missile still sitting on the rail. Just as doubt began to creep into my mind, the missile rocketed off my wingtip faster than anything I've ever seen before and disappeared from sight. I heard the whoosh and smelled the rocket propellant, but

neither registered as I marvelled at the speed with which the missile left my aircraft. Watching the tapes during debrief showed no delay between the pressing of the 'pickle button' and the missile leaving the aircraft.

As I fire my missile this time, the enemy Tiger pukes out a flare and executes a break turn, possibly defeating my shot. I choose to follow-up my missile shot with the gun. Slamming the throttle into AB, I fly towards the spot where the Tiger initiated his break turn, roll to the left, and pull back on the stick while blending in the pressure until I hit 9g.

OK, I've over-g'd the target pod in this case, but I wanted to tell you about pulling 9g. I've already taken a big prep breath and squeezed my legs prior to executing my turn. It's best to be on top of the AGSM before the actual g-onset. As the gs increase, I set my head on the back of the headrest. The human head weighs 10–12lb and the JHMCS weighs another 1.5lb. So at 9g my head weighs over 110lb. If I don't rest it against the seat at high g, I'm likely to injure myself. I continue to pull the Tiger closer to my HUD as I strain to keep the blood in my head. Relaxing on the pull, I drop down to 6g and patiently wait until I see my gun attack cues. I want to see roughly 45° of aspect, which is when the enemy jet appears to have the same width as it does length, and have closed the range between us to about 4,000ft. A common problem early on in training, when learning to employ the gun, is pulling the enemy into the HUD for a gunshot too soon. Being patient and aggressive at the same time can be tough to master. It takes patience to wait until aspect and range are correct and closure is under control. If I pull too soon, I will have a fleeting shot and possibly overshoot the Tiger, driving my offensive position to a neutral position or, even worse, defensive.

Seeing my attack cues, I pull the Tiger aggressively into my gun funnel, slamming the throttle into idle and thumbing my speed brakes out to slow my rate of closure. I continue to pull until he is in the right spot for me to shoot. As I squeeze the trigger, the Tiger jinks, spoiling my shot and creating a massive amount of closure. I pull him out of my HUD and roll around him arresting the closure rate. Once the closure is under control, I again pull him rapidly into my gun funnel. He has significantly reduced his energy with the jink and won't be able to manoeuvre as aggressively any more. I establish the appropriate amount of lead, squeeze the trigger, achieve a valid gun track and call the kill.

Shooting the gun is awesome! I've not yet had the pleasure of firing live at an aerial target, but I've made countless strafe passes at ground targets. The M61 Vulcan is nestled slightly behind and to the left of me. When fired, the gun is loud and rattles the cockpit, sometimes causing insulation behind the interior panels to fly through the cockpit. I admit the first time I shot the gun it startled me, and my strafe pass was terrible. Ever since then it's been a reassuring growl that spews 100 20mm bullets per second. Even more spectacular is firing the gun at night and seeing the flame of bullets fly out of the cannon. Through my NVGs it looks like hundreds of bottle rockets being launched at once.

Two and I continue our bug out towards friendly territory, clearing the airspace ahead with radar, while listening on the radio for Three and Four's progress and any updates from AWACS on enemy activity. Three and Four have successfully killed their targets and we work to rejoin as a four-ship while hauling-ass out of enemy airspace. I see no contacts on my radar and AWACS confirms there are no aircraft impeding our egress. As we cross over into friendly territory, I slow the formation down and we FENCE-out, turning our offensive and defensive systems to safe, stow the target pod and put the jet back into NAV mode. Everyone is slowly retracting their fangs as we start our way back home. I check out of the airspace with AWACS and proceed to RTB.

RTB

I have the four-ship rejoin to close formation, slow to 300kt, and conduct a battle damage (BD) check. A BD check is a visual lookover of each other's aircraft for any abnormalities, whether it is actual damage, an opened panel, a hung-bomb or fluid leaks. Anything out of the normal is checked for. I conduct another fuel check, which I've been doing periodically throughout the sortie. I push the flight out to Fluid-Four formation – Three flies line abreast one mile off of me and the wingmen fly Fighting

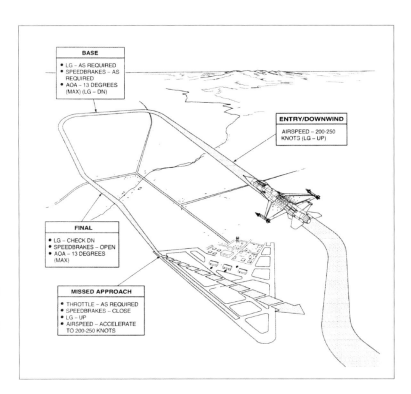

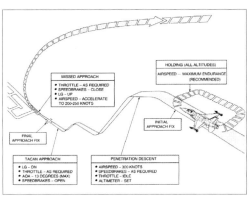

Returning to base, several options are available: a visual approach, a ground-controlled intercept approach using radar vectors [above] and a TACAN (tactical air navigation) approach. *(USAF)*

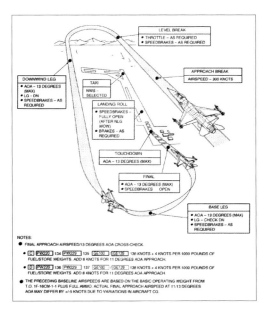

TOP AND ABOVE
Keeping a tight formation while flying up Initial is seen as an important part of the recovery. From the ground, seeing either a two-ship or four-ship of Vipers break overhead is an impressive sight. However, it is also important in creating the necessary spacing during landing so that as many aircraft can be recovered as quickly and as safely as possible in a short space of time.
*(Steve Davies/
FJPhotography.com)*

Wing (500ft and 30–70°) off their respective Flight Lead's wings.

We continue to scan our radars and listen to the radio for traffic conflicts now, not enemy targets. Now that the busy (and fun) part is over, I watch myself and the flight for complacency. Though I'm not near as busy now as I was a few minutes ago, I have to remember that I'm still flying and must continue with my crosscheck.

Landing

Approaching the base for landing, I have a few options. If the weather is bad, we fly an instrument approach in radar trail. The Flight Lead uses his radar to clear for the flight and flies the vectors given by radar approach. Two, Three and Four are each saddled two miles astern of the preceding aircraft, using their radar to maintain position. The only reason I would choose an instrument approach is for weather

or training requirements. Otherwise, I'm flying up Initial.

Initial is generally 1,500ft above ground level (AGL), three miles extended centreline from the runway, and the initial part of a visual flight rule (VFR) pattern. We fly at 300kt in four-ship fingertip formation. We arrive over the airfield looking sharp and conduct our break turns at five-second intervals to inside downwind.

I pull 3–4g in my level break turn, pull my throttle back to 80% rpm and put out my speed brakes. Rolling out, I check my spacing from the runway, guarantee I'm around 240–220kt, and note good hydraulic pressures. I lower my gear, check that I get three green indications, and continue proceeding to the perch, adjusting my speed to arrive at 200kt. At the perch, I lower my nose and bank, beginning my base to final turn while calling on the radio 'Viper One, base gear, stop.' I receive clearance to land and continue my turning descent.

I establish my aim point a mile from the runway by lining up on a 2.5° glideslope and 2.5° line in the HUD, FPM (flight path marker), and threshold of the runway. I'm roughly 160kt (depending on fuel load) 11° AoA, flying toward the runway. Crossing the threshold, I shift my aim point down the runway and slowly pull my throttle back holding 11–13° AoA and fly the jet to the ground.

On touching down, I hold the Viper's attitude at 13° to dissipate as much airspeed as possible without use of brakes. At 100kt I lower the nose to the ground and apply steady brakes, pulling my speedbrake switch out to full open, and slow the aircraft down. Reaching a safe taxi speed, I select NWS and taxi clear of the runway. The Viper is a relatively easy aircraft to land and any landing I can walk away from is a good landing.

I taxi into EOR, where the de-arming crew checks my aircraft for anything unsafe and pins my chaff and flares. We taxi back to our parking spots as a four-ship. My crew chief marshals me into my spot, and I ready the aircraft for shutdown. He hooks up on the intercom and he and his 'B man' check over the aircraft once again, noting pressures of systems that may need to be serviced and anything out of the ordinary. Once this is complete he tells me I'm clear for shutdown.

I turn off all my avionics, pull the throttle to

RIGHT The F-16 is landed at between 11° and 13° AoA, then held in this attitude to create as much aero braking as possible. Here, a 79th FS Block 50 pilot shows how it's done. *(USAF)*

off, and as the engine spools down so do my mind and body. I raise my canopy and unstrap the jet. The crew chief is at the ladder and I hand him my helmet bag prior to climbing down. I walk around the jet, looking for leaks, bird strikes or other damage by dragging or patting my hand on the metal surface, much like a jockey would a horse after a race. I'm once again humbled at how awesome it is to have flown this jet. I shake hands with the crew chief, thank him for letting me borrow his jet, and go inside to log my hours and prepare for debrief.

Debrief

Debrief is the bread and butter of being a good fighter pilot. This is where the majority of learning takes place. Rank, egos and excuses are left at the door. We will run through our tapes, validate our shots/bomb drops, listen to our communications, look at our avionics and see if everyone is sticking to the contracts or what was briefed. We see if our tactics are sound and how well we executed them, while identifying areas for improvement or change. Everyone admits to their mistakes and we can all learn from them.

One area of weakness from this mission that I would target is how I got jumped by the enemy F-16. Did we miss something? Is there something we can do to make sure it doesn't happen again? Nobody is perfect, but we continue to learn and we strive towards perfection together. Once debrief is over, we pick up our rank and egos at the door and head to the bar to tell our stories of conquest to the rest of our brothers and sisters.

Though flying the Viper is an absolutely awesome experience that I wouldn't trade for the world, it's the people that I fly, work and live with that I will miss most after my time in the F-16 is done. Their dedication to the mission and each other is unmatched, and I will always treasure the time spent with this family more than any time spent flying the Viper.

ABOVE Taxiing back to either a parking spot or a hardened aircraft shelter, weapons and sensors are turned off and, once the maintenance crew have completed some quick checks, the throttle is pulled all the way back into the cut-off position to kill fuel flow to the motor. *(USAF)*

BELOW A Viper pilot endures the rain as he inserts a removable data cartridge into an air combat manoeuvring and instrumentation (ACMI) pod. Such pods record aircraft navigation, weapons and targeting data, allowing the pilot to analyse and learn from each sortie. *(USAF)*

Chapter Eight

Maintaining the Viper: the technician's view

Josh Smith, a former F-16 flight line maintainer and now an F-16 pilot, reflects in this chapter on what it takes to maintain the Viper.

OPPOSITE Battered by ice, snow, sand, wind, rain; exposed to sub-zero temperatures and roasting sun; the maintainer is without question the unsung hero of the story of air power, past and present. These men and women, about whom little is ever said, work magic. *(USAF)*

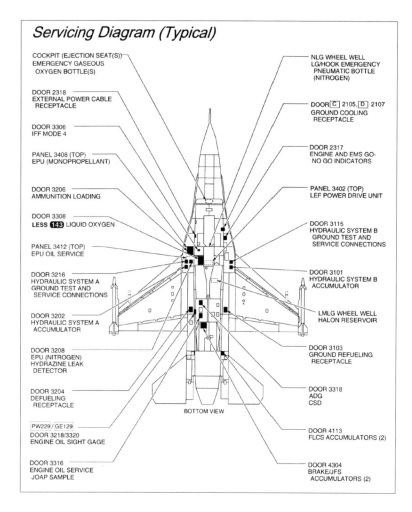

Servicing Diagram (Typical)

COCKPIT (EJECTION SEAT(S))
EMERGENCY GASEOUS
OXYGEN BOTTLE(S)

DOOR 2318
EXTERNAL POWER CABLE
RECEPTACLE

DOOR 3306
IFF MODE 4

PANEL 3408 (TOP)
EPU (MONOPROPELLANT)

DOOR 3206
AMMUNITION LOADING

DOOR 3308
LESS 143 LIQUID OXYGEN

PANEL 3412 (TOP)
EPU OIL SERVICE

DOOR 3216
HYDRAULIC SYSTEM A
GROUND TEST AND
SERVICE CONNECTIONS

DOOR 3202
HYDRAULIC SYSTEM A
ACCUMULATOR

DOOR 3208
EPU (NITROGEN)
HYDRAZINE LEAK
DETECTOR

DOOR 3204
DEFUELING
RECEPTACLE

PW229 / GE129
DOOR 3218/3320
ENGINE OIL SIGHT GAGE

DOOR 3316
ENGINE OIL SERVICE
JOAP SAMPLE

NLG WHEEL WELL
LG/HOOK EMERGENCY
PNEUMATIC BOTTLE
(NITROGEN)

DOOR C 2105, D 2107
GROUND COOLING
RECEPTACLE

DOOR 2317
ENGINE AND EMS GO-
NO GO INDICATORS

PANEL 3402 (TOP)
LEF POWER DRIVE UNIT

DOOR 3115
HYDRAULIC SYSTEM B
GROUND TEST AND
SERVICE CONNECTIONS

DOOR 3101
HYDRAULIC SYSTEM B
ACCUMULATOR

LMLG WHEEL WELL
HALON RESERVOIR

DOOR 3103
GROUND REFUELING
RECEPTACLE

DOOR 3318
ADG
CSD

DOOR 4113
FLCS ACCUMULATORS (2)

DOOR 4304
BRAKE/JFS
ACCUMULATORS (2)

BOTTOM VIEW

BELOW A crew chief assigned to the 51st Aircraft Maintenance Squadron references an electronic technical order, or TO. TOs provide reference and procedural instructions for Viper maintainers. Running to many volumes, their conversion to electronic format has made life just a little easier. *(USAF)*

Over the past ten-plus years I have had the opportunity to work two different airframes, both while serving on Active Duty and while being a member of the Air National Guard. I spent four years, directly out of high school, working as a Block 50 F-16CJ Flight Line Avionics Technician as a member of the 55th Aircraft Maintenance Unit (AMU) at Shaw AFB, South Carolina. Following that I spent four years working as a KC-135R Communication and Navigation Technician for the 186th ARW, Mississippi Air National Guard, at Meridian AFB, and then went to work Block 42 F-16s with the 180th FW, Ohio Air National Guard, for two years before being selected to fly the F-16 for the Minnesota ANG. Regardless of airframe or branch of service, one thought keeps recurring in my mind: what happens on the flight line is truly amazing.

Murphy's Law is standard operating procedure on the flight line. It's expected by all. It is a dangerous and dynamic place, but what happens there ultimately decides whether or not we, as an Air Force, can fully execute our mission. The days are long and tiring, both physically and mentally. Work is constant, day and night, through all weather conditions, doing whatever it takes to make sure enough jets are greened up to fly the next day's sorties. Sometimes that means getting called in to work in the middle of the night, after a long day's work already, because the unexpected happened and we need to pool resources to get a jet fixed. Other times that means being the only individuals driving towards the base – while the entire State is evacuating before a large hurricane hits – so that we can launch our jets to safety or get them inside the hangars. And one of the truly incredible things about all this is that the majority of the technicians doing the actual work on the flight line are 18–22-year-old kids.

When I was 18 years old I was in tech school learning how to work on the Viper. Two years later, I was a subject matter expert for anything avionics-related (and very well versed on many other areas of the jet), as were many of my friends. At that time, I was an Airman First Class (A1C), and had two stripes on my sleeve. We were it. We were the ones fixing the hard broke jets and figuring out the highly complicated problems that were intrinsic to such a high-tech

RIGHT The flight line is a dangerous and dynamic place. Here, a crew chief ducks under the wing of an Aggressor Viper – no longer a static object, the F-16 has become a living, breathing entity that can cause severe injury, or even death, to those who do not keep their wits about them. *(USAF)*

CENTRE When lucky, maintainers can work under the cover of a hardened aircraft shelter (HAS) as shown here. Even then, though it will protect them from most of the elements it might not protect them from intense heat or severe cold. When unlucky, maintainers must work entirely exposed to the weather. *(USAF)*

machine. We were trained by the best NCOs on the line, who would eventually get orders and leave; we were expected to fill their shoes. It's a sink-or-swim deal, though, because jets have to fly and peoples' lives are on the line. Some rose to the top while others were weeded out quickly, tasked to do other jobs on the line in support of those working on the planes.

Launch/EOR/recovery/ hydrazine

Launching jets is a well-orchestrated event, and with good reason. It's the business end of what we do: putting metal in the air. There are pretty much three parts to a launch. The first is getting the jet cranked and ready to taxi. This is a fairly standard procedure and typically will consist of a crew chief as the A man, on the comm cord with the pilot, and either an avionics or weapons guy as the B man, standing fireguard for the JFS when the engine cranks, as well as assisting in pulling pins and chalks. As an avionics troop, this is one of the first things you need to learn and will be responsible for. While pretty standard, this is also a pretty dangerous time, even for those who are well versed in the dangers. The engine is running, which gives you caution zones around the intake and exhaust. The flight control surfaces are moving, and – with 3,100psi behind the hydraulics – they create a serious area of concern. Also, the chord line of the wing sits about eye-level off the ground, which makes

BELOW Airmen assigned to the 169th Fighter Wing prepare to launch the Block 52 Vipers of the SCANG (South Carolina ANG). For their efforts, a lucky few F-16 maintainers will be given an 'incentive ride' – a flight in the F-16 – as a gesture of thanks and as a way of allowing them to experience first-hand the importance of their work. *(USAF)*

137

THIS PAGE Launching the Viper requires a well-orchestrated performance from at least two F-16 maintainers, each of whom has distinct responsibilities that must be performed in the correct sequence and at the correct time. All the while, they must do so in an extremely hazardous environment – jet exhaust, EPU fire, moving flight control surfaces and a range of other dangers are constantly present. *(USAF)*

it easy to bump your head against if you're lacking situational awareness.

The second part of the launch is the 'redball' portion. Redballs are unplanned (though not unexpected) problems with the jet that show their face in the middle of a launch. Essentially, as the pilot is turning on systems and running through the checks, one (or more) of the systems reports an error or simply does not work. The pilot then calls the redball in over the radio and the proper truck (crew chiefs, weapons, or avionics, EE or engines specialists) responds by pulling up to the jet and dropping an experienced technician off to troubleshoot the problem on the spot. The goal is to get the problem fixed or inform the pilot of the severity of the issue without causing the engine to shut down, so that the jet can fly the sortie without stepping to a spare.

Not all problems keep a plane from flying; sometimes the mission can still be accomplished with a degraded or non-functioning system. Sometimes this is caused simply by pilot error (wrong inputs, wrong sequencing, etc), other times it's just a fluke and recycling system power (off, then on again) corrects the issue, but occasionally it's actually broken. If it's the last, there are a few options: first, if the maintainers know this problem as a common issue and are confident of a quick fix by swapping out a simple part, then that will be done on the spot, which sometimes means 'canning' (cannibalising) a part out of the non-flying jet in the next spot over and swapping it out with the jet that's flying. If the problem is more troublesome and perplexing, often a spare jet will be waiting and the pilot will shut down the current jet to step to the spare. Redballs are common, and a few jets will have them occur every single time a launch happens. It's just the nature of the beast.

The last part is pulling chocks, taxiing the jet out and getting it to EOR. At this point, it's almost a sure thing that the jet will have a successful sortie, but sometimes they get to EOR before a problem pops up. By the time the jet gets down to EOR they're very limited on the time they have left to take off, so if the problem cannot be fixed immediately the jet will be forced to turn around and come back to chalks with no step to the spare.

RIGHT Maintainers will have a speciality – avionics, weapons (shown here), engine etc. The crew chief career field requires more general knowledge, leading to the formal term APG (airplane, powerplant general) for the men and women assigned the responsibility of maintaining their own Viper. Launches are run by a crew chief, but the B man on the launch team is often a weapons specialist. *(USAF)*

EOR is simply a last set of eyes on the aircraft before it launches. Typically, a team for the day is assembled, usually consisting of technicians from each profession, and they spend the day down at EOR. EOR also serves as the final point for extra safety-pins to be pulled/re-pinned from weapons pylons and what not. Really not a whole lot to say here.

The final phase of a successful sortie is the recovery. Usually, about 15–20 minutes out from base, the jets call in with their squawks, informing the maintainers of the condition of the jet before it lands. This allows us to get a specialist out to the ramp and at the correct spot so that when the jet rolls into chalks we can start troubleshooting the problem. Getting to speak with the pilot for a few minutes about exactly what happened and having them try a few things before the engine shuts down is a critical step in the troubleshooting process. Typically, each jet will squawk with one of four codes: Code 1, Code 2, Code 3 and IFE (in-flight emergency).

Code 1 means the jet is good to go: shut the engine down, perform the post-flight inspections and get gas in it for the next sorties. Code 2 means there is a degrade of some sort. Again, this doesn't necessarily mean the jet cannot fly again without the problem being fixed, but we typically try to fix it before

RIGHT Returning Vipers will pass their 'code' status over a discrete squadron frequency about 15 minutes prior to arrival. This heads-up allows the scheduling shop and the maintainers to plan ahead for the rest of the day's sorties. In this image, crew chief and assistant crew chief recover 'their' F-16 after it has completed its 30th Code 1 flight. *(USAF)*

LEFT Redballs – faults that may prevent an aircraft from completing its mission – can sometimes be solved by replacing a 'black box'. More problematic faults may result in the Viper being shut down and towed to a hangar for more thorough inspection. Here, an engine specialist opens panels to check an engine fault. *(USAF)*

the next set of goes (time permitting). Code 3 means the jet is broken and will not fly until the discrepancy is corrected. This is typically a complete system failure and requires in-depth troubleshooting. If a jet comes back Code 3 from the first set of goes for the day, it will most likely be pulled out of the line-up and replaced with a spare. Then there are the IFEs, which

are rare but occur when something extremely abnormal occurs during the flight. Sometimes it is a system failure and the pilot is required to call in the IFE, like a dual flight control failure. Other times it can be numerous things: engine out, bird strike, canopy seal failure (blowing the pilot's eardrums), engine fire, uncommanded flight control manoeuvre, etc.

If the jet squawks a Code 1–3 it will make a stop at EOR and get pinned up (if need be), then taxi back to chalks. Similar to the launch, an A man and B man will be in position waiting. A few extra hazards exist here, namely hot brakes and static electricity build-up on the jet. The A man typically marshals the jet into the spot, stopping it early so that we can check the condition of the tyres and for FOD (foreign object damage), and so that the B man can check the brakes to ensure they aren't too hot. If the brakes are good, the A man marshals the jet into its final position and the ground chord is connected to the ground, then to the aircraft to provide a path for electrostatic discharge. Then we chalk/pin the aircraft up (main landing gear pins, EPU, external fuel tanks) and continue standard checks before shutting the engine down. Hot brakes are a scary thing because, as they sit, they get hotter (lack of airflow across the metal). During this time they are radiating heat out to the tyre, and if the temperatures get too high the tyre will blow. There is a significant amount of pressure in that tyre and, if it blows the whole hub assembly is going to come off with it. Essentially it would be like shooting a gigantic cannonball across the flight line; anything in its path would be destroyed.

At this point the engine is ready to be shut down and the sortie is complete. This is the only real time a threat of hydrazine is present. If the engine is shut down before the emergency power unit (which uses hydrazine for fuel) gets pinned by the crew chief, then the EPU will fire and anyone downwind of the aircraft will be exposed to hydrazine. The EPU makes a loud, whining, turbo-like sound, and it's fairly obvious when it fires. There is also an indicator to check on the exterior that will let you know if the EPU has fired or not. If the EPU did fire, the personnel on the line will have to begin the decontamination process and then be checked out by medical – not a fun thing, because it involves taking damn near all your clothes off on the flight line (to avoid spreading the contamination). In my experience it is very rare to see the EPU fire – the odds are almost non-existent.

If everything goes well on the shutdown, the pilot gets out, the crew chief pins the ejection seat before anyone else can touch it, and we begin the post-flight process. The crew chief and engines troops will jump the tubes (intake and exhaust inspection) and the pilot will write up any discrepancies in the aircraft forms, then head to debrief. The fuel trucks come and it's a mad dash to get the jets ready for the next sortie of the day. Sometimes, if it's a slow day, we may only have one or two sets of goes; a normal day is typically three, and anywhere between four and twelve jets per set. Once in a while we will surge, which means non-stop flying, and will sometimes handle five to seven sets of sorties, ending with 55–70 individual sorties for the day. That's when things really get busy and dangerous because,

in conjunction with so many aircraft moving on the ramp at once, we also refuel them without shutting the engine down (commonly referred to as 'hot pitting').

There is one other possible outcome from a sortie for which no code is squawked. The jets start squawking and landing, pulling into chalks, and the realisation hits that one or two jets haven't squawked. Confusion sets in for a minute and all the high-ranking individuals are gathering in the unit. Jets are shutting down and MPs begin to barricade the unit; nobody is coming in or going out. A team comes in, snatching up all the aircraft forms and shop log books (used to document notes for shift change), and a bus pulls up for the maintainers to get on to go take a urine analysis test. By this time most of us have figured it out: one of our jets is not coming home.

The most sombre feeling overwhelms the unit, and prayers for the pilot's safety now bombard your thoughts, intermixed with concern that you may have missed something while you were working on the jet the night prior, and you may have been the cause of the crash. The chief gathers everyone and briefs us that one of our own went down and lets us know what to expect for the next few days. PR personnel brief us about how to handle the press and the public, while all flying comes to a halt for days to come. Shortly after, the family of the downed pilot arrives at the unit – the spouse in tears, not knowing the condition of their loved one, and the children young enough to not realise what's going on. At this point you can't help but think to yourself 'What if I

screwed up? What if I killed her husband and their father?' It's a truly horrible feeling, and one that haunts the squadron for quite some time thereafter. I have experienced this twice in my career and, thankfully, all the pilots returned to flying after some sort of recovery process.

Work between sorties/ troubleshooting

An Aircraft Maintenance Unit usually runs 24/7, consisting of three shifts: day shift, swing shift (evening), and mid shift (starts at midnight or so and runs until day shift comes back in). Typically all the flying will be done on day shift, with the exception of a week every now and then when we night fly. As a day shifter, you can expect to launch and recover jets and that's about it, reason being that there is usually only an hour and a half to two hours between sorties, which just isn't enough time to break into a broken jet. Launching jets requires damn near everybody in the unit, so you can only afford to have a few people working a jet while the rest are on the truck launching/recovering all the other aircraft. Another common task is to reconfigure jets for the next set of goes, which involves moving target, HTS (HARM targeting system) or ECM (electronic countermeasures) pods from one jet to another. Other than that, the shift that flying occurs on is typically overwhelmed with the sorties and usually does not do much in-depth troubleshooting.

Swing shift will come in and relieve day shift. Typically this is when the hardcore troubleshooting starts. The jets usually are not flying, so we have

LEFT A pair of Block 42Gs of the 180th Fighter Wing, Ohio ANG, await the start of inspections. Note the use of red padded covers for the trailing-edge flaps. These prevent head injuries for unwary maintainers who may walk into the eye-level devices. *(USAF)*

LEFT As the day wears on, the night shift will take over. Keeping the F-16 flying requires a constant stream of experts working in shift patterns. However, when sorties must be generated the shift pattern may go out the window. This Viper crew chief walks the wing inspecting the leading-edge flaps for faults or signs of defect. *(USAF)*

from early afternoon until the next morning to get the jets fixed and back in the line-up for the next day. This is typically the shift where your most experienced personnel are going to be. Jets get torn apart, parts are changed, new wires are run, and all the required follow-on maintenance and ops checks are performed. Sometimes the part is replaced and the jet is greened up within a few hours, sometimes days are spent trying to figure out what the problem is. I'd say probably 80% of the time we can figure the problem out and get it fixed that evening, but for the other 20% of the time all the manuals and engineers are out of ideas and you just have to really start thinking outside the box to find the problem. This is actually quite

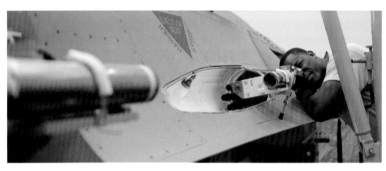

ABOVE LEFT A weapons specialist looks through a boresight telescope on an F-16 trainer device to ensure that the M61A1 Vulcan cannon barrels are correctly aligned. As he watches the crosshairs of the scope, another airman moves a collimator (foreground) so the weapon alignment is accurate. *(USAF)*

LEFT A 510th Aircraft Maintenance Unit avionics specialist tests the lower radar warning receiver antennae on an F-16C Block 40 at Aviano Air Base, Italy. Increasing computerisation throughout the F-16's lifetime has increased both the number of, and sophistication of, the tests that flight line avionics specialists can now undertake. *(USAF)*

FAR LEFT Resting on the port external fuel tank, a crew chief selects a safety clip during repair work of the leading edge flap. *(USAF)*

an art form and is what really separates a good maintainer from a great maintainer.

Troubleshooting typically starts with the easiest, most common solution, following a logical fault tree in one of the technical orders, from replacing one part to the next, working our way into the least likely scenario. To do this, though, you have to have a good understanding of what the problem is, which is why getting to speak with the pilot as soon as they come back from the sortie is so crucial. Even then, a maintainer will typically jump in the cockpit and run up the system that's in question, trying to duplicate the problem, and then manipulate some inputs to help gain a better understanding of it. Sometimes the first thing you try fixes the jet, and other times you do everything the book tells you to do and the problem still exists. At this point you really have to dive back into your knowledge of the theoretical system operation and how it interacts with all the other systems on board. That's one of the main problems with such a highly complicated jet: the systems are so highly integrated that is takes a very sound knowledge base to understand how they interact, and it isn't uncommon for one system to cause another to fail. Sometimes it's a bad wire in the system and you have to break out the wiring diagrams, start making educated guesses about what could be going on, and then check wires that carry power/ground/signal to find the culprit. All the while, you're up against the clock to get this jet fixed and greened up so it can fly.

Mid shift comes in around midnight and relieves swing shift. Mid shift will usually finish up anything the swings didn't have time to complete, then work on making sure all the jets are tidied up and ready for the morning sorties. This typically involves re-keying the jets with new codes for the day or month, depending on the system, crew chiefs servicing the aircraft,

and then a few will be spared to work on a jet that is still broken and was turned over from swings. Mid shift is typically a pretty small crew with very little supervision, which sometimes makes for a fairly laid-back atmosphere.

Every so many hours the jet needs to go into Phase and have some routine maintenance and inspections done. This involves tearing the jet down to nothing and inspecting/servicing multiple areas. Each Phase is dependent on

ABOVE LEFT A maintainer prepares to refuel an F-16 during RED FLAG-Alaska. Alaska provides a spectacular setting for the exercise, but the winter months force maintainers to endure weeks of sub-zero temperatures. *(USAF)*

LEFT The front fuselage fuel tank provides just enough space for a slender maintainer to crawl inside. Working with fuel tanks – or, more specifically, within the fume-filled environment of a fuel tank – is a dangerous business. Excessive fumes can lead to incapacitation and even death, necessitating that such work is always done in teams of two. *(USAF)*

LEFT Two 332nd Expeditionary Aircraft Maintenance Squadron crew chiefs test an LEF actuator during a Phase inspection at Balad AB, Iraq. Phase involves a more complete tear-down of the Viper than the usual launch and recovery inspections and servicing. *(USAF)*

RIGHT AND FAR RIGHT When a Viper reaches the end of its service life, one of several fates awaits it. It may become a ground instructional airframe (GF-16), or, as seen here, it may be sent to the AMARC boneyard, where contractors prepare it for long-term storage in the baking desert. The ill-fated aircraft in these images is an F-16A Block 15R. It may be years before its final destination is known, but it may well end its life as a target drone. *(USAF)*

either operating hour intervals or calendar days. Usually a Phase jet is going to be down for three to four weeks, at which point it is returned to the line. The Phase dock at each base typically consist of avionics, weapons, engines, crew chiefs etc from each unit, and they will work the next Phase or two down there, completely removed from the operational flight line and the day-to-day flying. The stuff they do in Phase is routine, so typically there are no surprises and no troubleshooting is required.

Other thoughts

One of the most amazing things about the flight line is the brotherhood and the camaraderie. The line is a harsh place to work. The hours are seriously long (a 12-hour day being a good day), and the weather can be brutal – sometimes it's so cold outside that we have to take turns working on the jet, each team only being able to stay out there for 15 minutes or so before having to come in (Air Force mandated), for fear of frostbite and hypothermia setting in. Sometimes it is so hot that merely touching your skin against the metal of the airframe will burn the hell out of you, leaving you with a blistered reminder to not do that again. Sometimes it means standing in the

freezing rain for two hours, waiting for the GPS to start tracking during a cloudy day, just to get a pass on the phase one inspection.

The only thing that matters on the flight line is this: jets have to fly. Going through these struggles and times of hardship together can really bring people close. I can't tell you how many times I was stuck working a hard broke jet when all my friends, who were working other jets, had finished their jobs and were told to go home because the shift was over. But instead of going home, they stayed, and came out to help me. Not because they thought I couldn't do it on my own, but because we were a team and we stuck together.

Another crazy thing, which I alluded to earlier, is the amount of responsibility that these young airmen accept, and how much we depend on them. I distinctly remember, as a two-striper, putting my foot down as my chief, ops squadron commander, the pilot and the production superintendent tried to get me to sign off the forms for a jet that was broken and was still in the line-up for the afternoon sortie. We had about 20 minutes until engine start time, and I wouldn't do it. I did not feel that the jet was good to fly, and they felt the opposite. Maybe it was, maybe everything would have been OK, but it wasn't worth it. It wasn't worth the risk of

BELOW AND BELOW RIGHT Keeping aircraft clean and freshly painted is also a responsibility of maintenance specialists and those who work in the 'paint barn'. Here, a Shaw-based F-16C Block 50 is cleaned by a maintainer wearing protective clothing and safety equipment. Meanwhile, a member of the paint barn team stencils a 480 EFS marking on to the vertical tail of a freshly cleaned Block 50 Viper. *(USAF)*

crashing a plane just so we could get a gold star and a pat on the back for not missing a sortie that day. It wasn't worth betting the pilot's life on it. Sometimes the higher-ranking dudes get blinded by the stats and metrics, and lose sight of the big picture. That's where the airmen and the NCOs have to step in and tactfully remind them what's at stake. Essentially it boiled down to me telling them all to f*** off. Not the most politically correct thing to do, but my chief had my back and the jet did not fly until we fixed the discrepancy. I was 19 years old. Think about what you were doing when you were 19. Now think about all the college kids and what they are doing during their freshman year, as 19-year-olds. The level of professionalism these young airmen have is unparalleled.

In general, maintenance feels drastically underappreciated and undermanned. I don't blame the others for not understanding what we do. It's too much for the non-flight-line-minded to comprehend. When the rest of the base is shut down due to an ice storm, we go in. When tornado warnings are present, we are there. Day or night, rain or shine, maintenance is there to make the mission happen. It's frustrating to see the dude working at the gym, wiping down exercise equipment, knowing he is getting paid the same as you. You also know that he works an eight-hour day, gets an hour for lunch, and gets time during his duty day to do his PT. You can't blame the guy, he is doing what he signed up to do, and works in an environment that allows him those freedoms, but it's frustrating. Same goes for the people at the chow hall. It's frustrating when, after 12 hours of working non-stop and missing meals, you go to the chow hall a few minutes before they shut down and they won't serve you. It's time for them to go home; they work on a strict hourly schedule,

and they don't want to stay even a minute late to help out another airman. Same goes for Finance and all the other support squadrons. At the end of the day, the mission really can happen without the gym or the chow hall. It can't happen without maintenance, and we take a lot of pride in that.

We work the long hours, come in on the weekends and in the middle of the night, put up with the weather and deployments, all because we truly believe in the mission and that the mission cannot happen without us. Watching the jets fly, come back without the bombs you sent them up with, and seeing the effect we are having in wartime is a pretty special thing. It's not something everyone in the military gets to experience, seeing first-hand their hard work put to use in a way that allows us to destroy the enemy and carry out our objectives. It is quite an amazing feeling when the pilots return to base, slick winged from all the ordinance they dropped, and you get them into chalks to shut them down. They get out, high fives and cheers all around, and you turn on the TV later that day to see the aftermath of the wrath our jets laid down. None of the long days, long deployments, sleepless nights, missed anniversaries/birthdays matter at that point. For that brief moment in time, all the sacrifice was worth it.

Appendix 1

Production table – production standard aircraft

Note: *The following data is correct to Summer 2013, and is provided courtesy of F16.net, the premier internet Viper resource.*

FY start	FY end	Country	Local s/n start	Local s/n end	No.	Comments
YF-16						
72-1567	72-1568	USA	01567	01568	2	
F-16A						
75-0745	75-0750	USA	75745	75750	6	1 converted to F-16XL/A
						1 converted to F-16XL/B
						1 converted to NF-16A (AFTI test bed)
						1 used as F-16/101 test bed
F-16B						
75-0751	75-0752	USA	75751	75752	2	1 used as F-16/79 test bed
F-16A Block 1						
78-0001	78-0021	USA	78001	78021	21	
78-0116	78-0132	Belgium	FA-01	FA-17	17	
78-0174	78-0176	Denmark	E-174	E-176	3	2 converted to F-16AM
78-0212	78-0223	Netherlands	J-212	J-223	12	
78-0272	78-0274	Norway	272	274	3	2 converted to F-16AM
F-16B Block 1						
78-0077	78-0098	USA	78077	78098	22	
78-0162	78-0167	Belgium	FB-01	FB-06	6	4 converted to F-16BM
78-0204	78-0205	Denmark	ET-204	ET-205	2	1 converted to F-16BM
78-0259	78-0264	Netherlands	J-259	J-264	6	
78-0301	78-0302	Norway	301	302	2	1 converted to F-16BM
F-16A Block 5						
78-0022	78-0076	USA	78022	78076	55	
78-0133	78-0140	Belgium	FA-18	FA-25	8	
78-0177	78-0188	Denmark	E-177	E-188	12	7 converted to F-16AM
78-0224	78-0237	Netherlands	J-224	J-237	14	
78-0275	78-0284	Norway	275	284	10	7 converted to F-16AM
78-0308	78-0325	Israel	100	138	18	Local s/n are not sequential
79-0288		USA	79288		1	
F-16B Block 5						
78-0099	78-0115	USA	78099	78115	17	
78-0168	78-0171	Belgium	FB-07	FB-10	4	3 converted to F-16BM
78-0206	78-0208	Denmark	ET-206	ET-208	3	3 converted to F-16BM
78-0265	78-0266	Netherlands	J-265	J-266	2	
78-0303	78-0304	Norway	303	304	2	1 converted to F-16BM
78-0355	78-0362	Israel	001	017	8	Local s/n are not sequential
79-0410	79-0419	USA	79410	79419	10	
F-16A Block 10						
78-0141	78-0161	Belgium	FA-26	FA-46	21	
78-0189	78-0203	Denmark	E-189	E-203	15	14 converted to F-16AM
78-0238	78-0257	Netherlands	J-238	J-257	20	5 converted to F-16AM
78-0285	78-0299	Norway	285	299	15	11 converted to F-16AM
78-0326	78-0354	Israel	219	264	29	Local s/n are not sequential
79-0289	79-0409	USA	79289	79409	121	
80-0474	80-0540	USA	80474	80540	67	
80-0649	80-0668	Israel	265	299	20	Local s/n are not sequential
80-3538	80-3546	Belgium	FA-47	FA-55	9	

FY start	FY end	Country	Local s/n start	Local s/n end	No.	Comments
F-16B Block 10						
78-0172	78-0173	Belgium	FB-11	FB-12	2	1 converted to F-16BM
78-0209	78-0211	Denmark	ET-209	ET-211	3	1 converted to F-16BM
78-0267	78-0271	Netherlands	J-267	J-271	5	3 converted to F-16BM
78-0305	78-0307	Norway	305	307	3	2 converted to F-16BM
79-0420	79-0432	USA	79420	79432	13	
80-0623	80-0634	USA	80623	80634	12	
F-16A Block 15						
78-0258		Netherlands	J-258		1	1 converted to F-16AM
78-0300		Norway	300		1	
80-0541	80-0622	USA	80541	80622	82	60 converted to F-16A ADF
						1 converted to F-16AM
80-0639	80-0643	Egypt	9301	9305	5	
80-3547	80-3587	Belgium	FA-56	FA-96	41	34 converted to F-16AM
80-3596	80-3611	Denmark	E-596	E-611	16	16 converted to F-16AM
80-3616	80-3648	Netherlands	J-616	J-648	33	25 converted to F-16AM
80-3658	80-3688	Norway	658	688	31	27 converted to F-16AM
81-0643	81-0661	Egypt	9306	9324	19	
81-0663	81-0811	USA	81663	81811	149	115 converted to F-16A ADF
81-0864	81-0881	Netherlands	J-864	J-881	18	16 converted to F-16AM
81-0899	81-0926	Pakistan	82701	85728	28	Local s/n are not sequential
						1 converted to F-16AM
82-0900	82-1025	USA	82900	821025	126	59 converted to F-16A ADF
82-1050	82-1052	Venezuela			3	Local s/n at random
82-1056	82-1065	Egypt	9325	9334	10	
83-1066	83-1117	USA	83066	83117	52	
83-1186	83-1188	Venezuela			3	Local s/n at random
83-1192	83-1207	Netherlands	J-192	J-207	16	14 converted to F-16AM
84-1346	84-1357	Venezuela			12	Local s/n at random
84-1358	84-1367	Netherlands	J-358	J-367	10	7 converted to F-16AM
85-0135	85-0140	Netherlands	J-135	J-140	6	5 converted to F-16AM
F-16B Block 15						
80-0635	80-0638	USA	80635	80638	4	2 converted to F-16B ADF
80-0644	80-0648	Egypt	9201	9205	5	
80-3588	80-3595	Belgium	FB-13	FB-20	8	6 converted to F-16BM
80-3612	80-3615	Denmark	ET-612	ET-615	4	4 converted to F-16BM
80-3649	80-3657	Netherlands	J-649	J-657	9	8 converted to F-16BM
80-3689	80-3693	Norway	689	693	5	5 converted to F-16BM
81-0662		Egypt	9206		1	
81-0812	81-0822	USA	81812	81822	11	5 converted to F-16B ADF
81-0882	81-0885	Netherlands	J-882	J-885	4	3 converted to F-16BM
81-0931	81-0938	Pakistan	82601	84608	8	Local s/n are not sequential
						1 converted to F-16BM
81-1504	81-1507	Pakistan	85609	86612	4	Local s/n are not sequential
82-1026	82-1042	USA	821026	821042	17	14 converted to F-16B ADF
82-1043		Egypt	9208		1	
82-1044	82-1049	USA	821044	821049	6	4 converted to F-16B ADF
82-1053	82-1055	Venezuela			3	Local s/n at random

FY start	FY end	Country	Local s/n start	Local s/n end	No.	Comments
83-1166	83-1173	USA	83166	83173	8	
83-1189	83-1191	Venezuela			3	Local s/n at random
83-1208	83-1211	Netherlands	J-208	J-211	4	4 converted to F-16BM
84-1368	84-1369	Netherlands	J-368	J-369	2	2 converted to F-16BM
F-16A Block 15 OCU						
85-0141	85-0146	Netherlands	J-141	J-146	6	6 converted to F-16AM
86-0054	86-0063	Netherlands	J-054	J-063	10	8 converted to F-16AM
86-0073	86-0077	Belgium	FA-97	FA-101	5	5 converted to F-16AM
86-0378		Thailand	10305		1	
87-0004	87-0008	Denmark	E-004	E-008	5	5 converted to F-16AM
87-0046	87-0056	Belgium	FA-102	FA-112	11	10 converted to F-16AM
87-0397	87-0400	Singapore	880	883	4	
87-0508	87-0516	Netherlands	J-508	J-516	9	9 converted to F-16AM
87-0710		Netherlands	J-710		1	
87-0713	87-0720	Indonesia	TS-1605	TS-1612	8	
88-0001	88-0012	Netherlands	J-001	J-012	12	10 converted to F-16AM
88-0016	88-0018	Denmark	E-016	E-018	3	3 converted to F-16AM
88-0038	88-0047	Belgium	FA-113	FA-122	10	9 converted to F-16AM
89-0001	89-0011	Belgium	FA-123	FA-133	11	11 converted to F-16AM
89-0013	89-0021	Netherlands	J-013	J-021	9	9 converted to F-16AM
90-0025	90-0027	Belgium	FA-134	FA-136	3	3 converted to F-16AM
90-0942	90-0947	Pakistan	91729	92734	6	Local s/n are not sequential
90-7020	90-7031	Thailand	40307	40318	12	
91-0062	91-0067	Thailand	10313	10318	6	
92-0404	92-0410	Pakistan	92735	93741	7	Local s/n are not sequential
93-0465	93-0481	Portugal	15101	15117	17	16 converted to F-16AM
F-16B Block 15 OCU						
86-0064	86-0065	Netherlands	J-064	J-065	2	2 converted to F-16BM
86-0197	86-0199	Denmark	ET-197	ET-199	3	3 converted to F-16BM
86-0379	86-0381	Thailand	10301	10303	3	
87-0001		Belgium	FB-21		1	1 converted to F-16BM
87-0022		Denmark	ET-022		1	1 converted to F-16BM
87-0066	87-0068	Netherlands	J-066	J-068	3	3 converted to F-16BM
87-0401	87-0404	Singapore	885	888	4	
87-0711	87-0712	Norway	711	712	2	1 converted to F-16BM
87-0721	87-0724	Indonesia	TS-1601	TS-1604	4	
88-0048	88-0049	Belgium	FB-22	FB-23	2	2 converted to F-16BM
89-0012		Belgium	FB-24		1	1 converted to F-16BM
90-0948	90-0952	Pakistan	91613	92617	5	Local s/n are not sequential
90-7032	90-7037	Thailand	40301	40306	6	
92-0452	92-0461	Pakistan	92618	95627	10	Local s/n are not sequential
93-0482	93-0484	Portugal	15118	15120	3	3 converted to F-16BM
F-16A Block 20						
93-0702	93-0821	Taiwan	6601	6720	120	
F-16B Block 20						
93-0822	93-0851	Taiwan	6801	6830	30	
F-16C Block 25						
83-1118	83-1165	USA	83118	83165	48	
84-1212	84-1318	USA	84212	84318	107	
84-1374	84-1395	USA	84374	84395	22	
85-1399		USA	85399		1	
85-1401		USA	85401		1	
85-1403	85-1407	USA	85403	85407	5	
85-1409		USA	85409		1	
85-1411		USA	85411		1	
85-1413		USA	85413		1	
85-1415	85-1421	USA	85415	85421	7	
85-1423		USA	85423		1	
85-1425		USA	85425		1	
85-1427		USA	85427		1	
85-1429	85-1431	USA	85429	85431	3	
85-1433		USA	85433		1	
85-1435		USA	85435		1	
85-1437		USA	85437		1	
85-1439		USA	85439		1	
85-1441		USA	85441		1	
85-1443		USA	85443		1	
85-1445		USA	85445		1	
85-1447		USA	85447		1	
F-16D Block 25						
83-1174	83-1185	USA	83174	83185	12	
84-1319	84-1331	USA	84319	84331	13	
84-1396	84-1397	USA	84396	84397	2	
85-1506	85-1508	USA	85506	85508	3	
85-1510		USA	85510		1	
85-1512		USA	85512		1	
85-1514	85-1516	USA	85514	85516	3	
F-16C Block 30						
85-1398		USA	85398		1	
85-1400		USA	85400		1	
85-1402		USA	85402		1	
85-1408		USA	85408		1	
85-1410		USA	85410		1	
85-1412		USA	85412		1	
85-1414		USA	85414		1	
85-1422		USA	85422		1	
85-1424		USA	85424		1	
85-1426		USA	85426		1	
85-1428		USA	85428		1	
85-1432		USA	85432		1	
85-1433		USA	85433		1	
85-1436		USA	85436		1	
85-1438		USA	85438		1	
85-1440		USA	85440		1	
85-1442		USA	85442		1	
85-1444		USA	85444		1	
85-1446		USA	85446		1	
85-1448	85-1505	USA	85448	85505	58	
85-1513		USA	85513		1	
85-1517		USA	85517		1	
85-1544	85-1570	USA	85544	85570	27	
86-0066	86-0072	Turkey	86-0066	86-0072	7	
86-0207	86-0209	USA	86207	86209	3	
86-0216		USA	86216		1	
86-0219		USA	86219		1	
86-0221	86-0235	USA	86221	86235	15	
86-0237		USA	86237		1	
86-0242	86-0249	USA	86242	86249	8	
86-0254	86-0255	USA	86254	86255	2	
86-0258	86-0268	USA	86258	86268	11	
86-0270		USA	86270		1	
86-0274	86-0278	USA	86274	86278	5	
86-0282		USA	86282		1	
86-0284		USA	86284		1	
86-0286	86-0290	USA	86286	86290	5	
86-0293	86-0295	USA	86293	86295	3	
86-0297	86-0298	USA	86297	86298	2	
86-0300	86-0371	USA	86300	86371	72	
86-1598	86-1612	Israel	301	337	15	Local s/n are not sequential
87-0009	87-0021	Turkey	87-0009	87-0021	13	
87-0217	87-0266	USA	87217	87266	50	
87-0268		USA	87268		1	
87-0270	87-0292	USA	87270	87292	23	
87-0294		USA	87294		1	
87-0296		USA	87296		1	
87-0298		USA	87298		1	
87-0300		USA	87300		1	
87-0302		USA	87302		1	
87-0304		USA	87304		1	

FY start	FY end	Country	Local s/n start	Local s/n end	No.	Comments
87-0306		USA	87306		1	
87-0308		USA	87308		1	
87-0310		USA	87310		1	
87-0312		USA	87312		1	
87-0314		USA	87314		1	
87-0316		USA	87316		1	
87-0318		USA	87318		1	
87-0320		USA	87320		1	
87-0322		USA	87322		1	
87-0324		USA	87324		1	
87-0326		USA	87326		1	
87-0328		USA	87328		1	
87-0330		USA	87330		1	
87-0332		USA	87332		1	
87-0334	87-0349	USA	87334	87349	16	
87-1661	87-1693	Israel	340	376	33	Local s/n are not sequential
88-0019	88-0032	Turkey	88-0019	88-0032	14	
88-0110	88-0143	Greece	110	143	34	
88-0397	88-0411	USA	88397	88411	15	
88-1709	88-1711	Israel	329	359	3	Local s/n are not sequential

F-16D Block 30

FY start	FY end	Country	Local s/n start	Local s/n end	No.	Comments
85-1509		USA	85509		1	
85-1511		USA	85511		1	
85-1571	85-1573	USA	85571	85573	3	
86-0043	86-0047	USA	86043	86047	5	
86-0049	86-0053	USA	86049	86053	5	
86-0191	86-0196	Turkey	86-0191	86-0196	6	
87-0002	87-0003	Turkey	87-0002	87-0003	2	
87-0363	87-0368	USA	87363	87368	6	
87-0370	87-0380	USA	87370	87380	11	
87-0382	87-0390	USA	87382	87390	9	
87-1694	87-1708	Israel	020	061	15	Local s/n are not sequential
88-0013		Turkey	88-0013		1	
88-0144	88-0149	Greece	144	149	6	
88-0150	88-0152	USA	88150	88152	3	
88-1712	88-1720	Israel	063	088	9	Local s/n are not sequential

F-16N Block 30

FY start	FY end	Country	Local s/n start	Local s/n end	No.	Comments
85-1369	85-1378	USA	163268	163277	10	
86-1684	86-1695	USA	163566	163577	10	

TF-16N Block 30

FY start	FY end	Country	Local s/n start	Local s/n end	No.	Comments
85-1379	85-1382	USA	163278	163281	4	

F-16C Block 32

FY start	FY end	Country	Local s/n start	Local s/n end	No.	Comments
84-1332	84-1339	Egypt	9501	9508	8	
85-1518	85-1543	Egypt	9509	9534	26	
85-1574	85-1583	South Korea	85-574	85-583	10	
86-0210	86-0215	USA	86210	86215	6	
86-0217	86-0218	USA	86217	86218	2	
86-0220		USA	86220		1	
86-0236		USA	86236		1	
86-0238	86-0241	USA	86238	86241	4	
86-0250	86-0253	USA	86250	86253	4	
86-0256	86-0257	USA	86256	86257	4	
86-0269		USA	86269		1	
86-0271	86-0273	USA	86271	86273	3	
86-0279	86-0281	USA	86279	86281	3	
86-0283		USA	86283		1	
86-0285		USA	86285		1	
86-0291	86-0292	USA	86291	86292	2	
86-0296		USA	86296		1	
86-0299		USA	86299		1	
86-1586	86-1597	South Korea	86-586	86-597	12	
87-0267		USA	87267		1	
87-0269		USA	87269		1	
87-0293		USA	87293		1	
87-0295		USA	87295		1	

FY start	FY end	Country	Local s/n start	Local s/n end	No.	Comments
87-0297		USA	87297		1	
87-0299		USA	87299		1	
87-0301		USA	87301		1	
87-0303		USA	87303		1	
87-0305		USA	87305		1	
87-0307		USA	87307		1	
87-0309		USA	87309		1	
87-0311		USA	87311		1	
87-0313		USA	87313		1	
87-0315		USA	87315		1	
87-0317		USA	87317		1	
87-0319		USA	87319		1	
87-0321		USA	87321		1	
87-0323		USA	87323		1	
87-0325		USA	87325		1	
87-0327		USA	87327		1	
87-0329		USA	87329		1	
87-0331		USA	87331		1	
87-0333		USA	87333		1	
87-1653	87-1660	South Korea	87-653	87-660	8	

F-16D Block 32

FY start	FY end	Country	Local s/n start	Local s/n end	No.	Comments
84-1340	84-1345	Egypt	9401	9406	6	
84-1370	84-1373	South Korea	84-370	84-373	4	
85-1584	85-1585	South Korea	85-584	85-585	2	
86-0039	86-0042	USA	86039	86042	3	
87-0369		USA	87369		1	
87-0381		USA	87381		1	
90-0938	90-0941	South Korea	90-938	90-941	4	

NF-16 Block 30

FY start	FY end	Country	Local s/n start	Local s/n end	No.	Comments
86-0048		USA	86048		1	

F-16C Block 40

FY start	FY end	Country	Local s/n start	Local s/n end	No.	Comments
87-0350	87-0355	USA	87350	87355	6	
87-0357		USA	87357		1	
87-0359		USA	87359		1	
88-0033	88-0037	Turkey	88-0033	88-0037	5	
88-0413		USA	88413		1	
88-0415	88-0416	USA	88415	88416	2	
88-0418	88-0419	USA	88418	88419	2	
88-0421	88-0422	USA	88421	88422	2	
88-0424	88-0426	USA	88424	88426	3	
88-0428	88-0433	USA	88428	88433	6	
88-0435	88-0441	USA	88435	88441	7	
88-0443	88-0444	USA	88443	88444	2	
88-0446	88-0447	USA	88446	88447	2	
88-0449	88-0450	USA	88449	88450	2	
88-0452	88-0454	USA	88452	88454	3	
88-0457		USA	88457		1	
88-0459	88-0460	USA	88459	88460	2	
88-0462	88-0463	USA	88462	88463	2	
88-0465	88-0468	USA	88465	88468	4	
88-0470	88-0471	USA	88470	88471	2	
88-0473	88-0474	USA	88473	88474	2	
88-0476	88-0477	USA	88476	88477	2	
88-0479	88-0480	USA	88479	88480	2	
88-0482	88-0483	USA	88482	88483	2	
88-0485	88-0486	USA	88485	88486	2	
88-0488	88-0489	USA	88488	88489	2	
88-0491	88-0492	USA	88491	88492	2	
88-0494	88-0495	USA	88494	88495	2	
88-0497	88-0498	USA	88497	88498	2	
88-0500	88-0501	USA	88500	88501	2	
88-0503	88-0504	USA	88503	88504	2	
88-0506	88-0507	USA	88506	88507	2	
88-0509	88-0510	USA	88509	88510	2	
88-0512	88-0513	USA	88512	88513	2	

FY start	FY end	Country	Local s/n start	Local s/n end	No.	Comments
88-0515	88-0516	USA	88515	88516	2	
88-0518	88-0519	USA	88518	88519	2	
88-0521	88-0523	USA	88521	88523	3	
88-0525	88-0526	USA	88525	88526	2	
88-0528	88-0529	USA	88528	88529	2	
88-0531	88-0533	USA	88531	88533	3	
88-0535	88-0538	USA	88535	88538	4	
88-0540	88-0541	USA	88540	88541	2	
88-0543	88-0544	USA	88543	88544	2	
88-0546	88-0547	USA	88546	88547	2	
88-0549	88-0550	USA	88549	88550	2	
89-0022	89-0041	Turkey	89-0022	89-0041	20	
89-0277		Israel	502		1	
89-0278	89-0279	Egypt	9901	9902	2	
89-2000	89-2001	USA	89000	89001	2	
89-2003		USA	89003		1	
89-2005	89-2006	USA	89005	89006	2	
89-2008	89-2009	USA	89008	89009	2	
89-2011		USA	89011		1	
89-2013	89-2016	USA	89013	89016	2	
89-2018		USA	89018		1	
89-2020	89-2021	USA	89020	89021	2	
89-2023	89-2024	USA	89023	89024	2	
89-2026	89-2027	USA	89026	89027	2	
89-2029	89-2030	USA	89029	89030	2	
89-2032	89-2033	USA	89032	89033	2	
89-2035	89-2036	USA	89035	89036	2	
89-2038	89-2039	USA	89038	89039	2	
89-2040	89-2044	USA	89040	89044	2	
89-2046	89-2047	USA	89046	89047	2	
89-2049	89-2050	USA	89049	89050	2	
89-2052		USA	89052		1	
89-2054	89-2055	USA	89054	89055	2	
89-2057	89-2058	USA	89057	89058	2	
89-2060	89-2069	USA	89060	89069	2	
89-2071	89-2072	USA	89071	89072	2	
89-2074	89-2075	USA	89074	89075	2	
89-2077	89-2078	USA	89077	89078	2	
89-2080	89-2081	USA	89081	89081	2	
89-2083	89-2084	USA	89084	89084	2	
89-2086	89-2087	USA	89087	89087	2	
89-2090		USA	89090		1	
89-2092	89-2093	USA	89092	89093	2	
89-2095	89-2096	USA	89095	89096	2	
89-2099		USA	89099		1	
89-2101	89-2102	USA	89101	89102	2	
89-2104	89-2105	USA	89104	89105	2	
89-2108		USA	89108		1	
89-2110	89-2111	USA	89110	89111	2	
89-2113		USA	89113		1	
89-2115	89-2116	USA	89115	89116	2	
89-2118	89-2119	USA	89118	89119	2	
89-2121	89-2122	USA	89121	89122	2	
89-2124	89-2125	USA	89124	89125	2	
89-2127		USA	89127		1	
89-2130	89-2131	USA	89130	89131	2	
89-2134	89-2134	USA	89134	89134	2	
89-2136	89-2137	USA	89136	89137	2	
89-2139	89-2140	USA	89139	89140	2	
89-2143	89-2144	USA	89143	89144	2	
89-2146	89-2147	USA	89146	89147	2	
89-2149	89-2150	USA	89149	89150	2	
89-2152	89-2153	USA	89152	89153	2	
90-0001	90-0021	Turkey	90-0001	90-0021	21	
90-0028	90-0035	Bahrain	101	115	8	Local s/n are not sequential

FY start	FY end	Country	Local s/n start	Local s/n end	No.	Comments
90-0703		USA	90703		1	
90-0709	90-0711	USA	90709	90711	3	
90-0714		USA	90714		1	
90-0717	90-0718	USA	90717	90718	2	
90-0723	90-0725	USA	90723	90725	3	
90-0733	90-0736	USA	90733	90736	4	
90-0742	90-0745	USA	90742	90745	4	
90-0753		USA	90753		1	
90-0756		USA	90756		1	
90-0763		USA	90763		1	
90-0771	90-0776	USA	90771	90776	6	
90-0850	90-0874	Israel	503	547	25	Local s/n are not sequential
90-0899	90-0930	Egypt	9903	9934	32	
90-0953		Egypt	9935		1	
91-0001	91-0021	Turkey	91-0001	91-0021	21	
91-0486	91-0489	Israel	551	558	4	Local s/n are not sequential
92-0001	92-0021	Turkey	92-0001	92-0021	21	
93-0001	93-0014	Turkey	93-0001	93-0014	14	
93-0485	93-0512	Egypt	9951	9978	28	
93-0525	93-0530	Egypt	9979	9984	6	
96-0086	96-0106	Egypt	9711	9731	21	
98-2012	98-2021	Bahrain	201	210	10	
99-0105	99-0116	Egypt	9732	9743	12	
F-16D Block 40						
87-0391	87-0393	USA	87391	87393	3	
88-0014	88-0015	Turkey	88-0014	88-0015	2	
88-0166		USA	88166		1	
88-0168		USA	88168		1	
88-0170	88-0171	USA	88170	88171	2	
88-0173	88-0174	USA	88173	88174	2	
89-0042	89-0045	Turkey	89-0042	89-0045	4	
89-2166		USA	89166		1	
89-2168	89-2169	USA	89168	89169	2	
89-2171	89-2174	USA	89171	89174	4	
89-2176		USA	89176		1	
89-2178		USA	89178		1	
90-0022	90-0024	Turkey	90-0022	90-0024	3	
90-0036	90-0039	Bahrain	150	156	4	Local s/n are not sequential
90-0777		USA	90777		1	
90-0779	90-0780	USA	90779	90780	2	
90-0782		USA	90782		1	
90-0784		USA	90784		1	
90-0791	90-0792	USA	90791	90792	2	
90-0794	90-0800	USA	90794	90800	7	
90-0875	90-0898	Israel	601	667	24	Local s/n are not sequential
90-0931	90-0937	Egypt	9801	9807	7	
90-0954	90-0958	Egypt	9808	9812	5	
91-0022	91-0024	Turkey	91-0022	91-0024	3	
91-0490	91-0495	Israel	673	687	6	Local s/n are not sequential
92-0022	92-0024	Turkey	92-0022	92-0024	3	
93-0513	93-0524	Egypt	9851	9862	12	
99-0117	99-0128	Egypt	9863	9874	12	
F-16C Block 42						
87-0356		USA	87356		1	
87-0358		USA	87358		1	
87-0360	87-0362	USA	87360	87362	3	
88-0412		USA	88412		1	
88-0414		USA	88414		1	
88-0417		USA	88417		1	
88-0420		USA	88420		1	
88-0423		USA	88423		1	
88-0427		USA	88427		1	
88-0434		USA	88434		1	
88-0442		USA	88442		1	
88-0445		USA	88445		1	

FY start	FY end	Country	Local s/n start	Local s/n end	No.	Comments
88-0448		USA	88448		1	
88-0451		USA	88451		1	
88-0455	88-0456	USA	88455	88456	2	
88-0458		USA	88458		1	
88-0461		USA	88461		1	
88-0464		USA	88464		1	
88-0469		USA	88469		1	
88-0472		USA	88472		1	
88-0475		USA	88475		1	
88-0478		USA	88478		1	
88-0481		USA	88481		1	
88-0484		USA	88484		1	
88-0487		USA	88487		1	
88-0490		USA	88490		1	
88-0493		USA	88493		1	
88-0496		USA	88496		1	
88-0499		USA	88499		1	
88-0502		USA	88502		1	
88-0505		USA	88505		1	
88-0508		USA	88508		1	
88-0511		USA	88511		1	
88-0514		USA	88514		1	
88-0517		USA	88517		1	
88-0520		USA	88520		1	
88-0524		USA	88524		1	
88-0527		USA	88527		1	
88-0530		USA	88530		1	
88-0534		USA	88534		1	
88-0539		USA	88539		1	
88-0542		USA	88542		1	
88-0545		USA	88545		1	
88-0548		USA	88548		1	
89-2002		USA	89002		1	
89-2004		USA	89004		1	
89-2007		USA	89007		1	
89-2010		USA	89010		1	
89-2012		USA	89012		1	
89-2017		USA	89017		1	
89-2019		USA	89019		1	
89-2022		USA	89022		1	
89-2025		USA	89025		1	
89-2028		USA	89028		1	
89-2031		USA	89031		1	
89-2034		USA	89034		1	
89-2037		USA	89037		1	
89-2040		USA	89040		1	
89-2045		USA	89045		1	
89-2048		USA	89048		1	
89-2051		USA	89051		1	
89-2053		USA	89053		1	
89-2056		USA	89056		1	
89-2059		USA	89059		1	
89-2070		USA	89070		1	
89-2073		USA	89073		1	
89-2076		USA	89076		1	
89-2079		USA	89079		1	
89-2082		USA	89082		1	
89-2085		USA	89085		1	
89-2088	89-2089	USA	89088	89089	2	
89-2091		USA	89091		1	
89-2094		USA	89094		1	
89-2097	89-2098	USA	89097	89098	2	
89-2100		USA	89100		1	
89-2103		USA	89103		1	
89-2106	89-2107	USA	89106	89107	2	

FY start	FY end	Country	Local s/n start	Local s/n end	No.	Comments
89-2109		USA	89109		1	
89-2112		USA	89112		1	
89-2114		USA	89114		1	
89-2117		USA	89117		1	
89-2120		USA	89120		1	
89-2123		USA	89123		1	
89-2126		USA	89126		1	
89-2128	89-2129	USA	89128	89129	2	
89-2132		USA	89132		1	
89-2135		USA	89135		1	
89-2138		USA	89138		1	
89-2141	89-2142	USA	89141	89142	2	
89-2145		USA	89145		1	
89-2148		USA	89148		1	
89-2151		USA	89151		1	
89-2154		USA	89154		1	
90-0700	90-0702	USA	90700	90702	3	
90-0704	90-0708	USA	90704	90708	5	
90-0712	90-0713	USA	90712	90713	2	
90-0715	90-0716	USA	90715	90716	2	
90-0719	90-0722	USA	90719	90722	4	
90-0726	90-0732	USA	90726	90732	7	
90-0737	90-0741	USA	90737	90741	5	
90-0746	90-0752	USA	90746	90752	7	
90-0754	90-0755	USA	90754	90755	2	
90-0757	90-0762	USA	90757	90762	6	
90-0764	90-0770	USA	90764	90770	8	
F-16D Block 42						
87-0394	87-0396	USA	87394	87396	3	
88-0153	88-0165	USA	88153	88165	13	
88-0167		USA	88167		1	
88-0169		USA	88169		1	
88-0172		USA	88172		1	
88-0175		USA	88175		1	
89-2155	89-2165	USA	89155	89165	11	
89-2167		USA	89167		1	
89-2170		USA	89170		1	
89-2175		USA	89175		1	
89-2177		USA	89177		1	
89-2179		USA	89179		1	
90-0778		USA	90778		1	
90-0781		USA	90781		1	
90-0783		USA	90783		1	
90-0785	90-0790	USA	90785	90790	6	
90-0793		USA	90793		1	
F-16C Block 50						
90-0801	90-0808	USA	90801	90808	8	
90-0810	90-0833	USA	90810	90833	24	
91-0336	91-0361	USA	91336	91361	26	
91-0363	91-0369	USA	91363	91369	7	
91-0371	91-0373	USA	91371	91373	3	
91-0375	91-0385	USA	91375	91385	11	
91-0387	91-0391	USA	91387	91391	5	
91-0402	91-0403	USA	91402	91403	2	
91-0405	91-0412	USA	91405	91412	8	
91-0414	91-0423	USA	91414	91423	10	
92-3883	92-3884	USA	92883	92884	2	
92-3886	92-3887	USA	92886	92887	2	
92-3891	92-3895	USA	92891	92895	5	
92-3897		USA	92897		1	
92-3900	92-3901	USA	92900	92901	2	
92-3904		USA	92904		1	
92-3906	92-3907	USA	92906	92907	2	
92-3910		USA	92910		1	
92-3912	92-3913	USA	92912	92913	2	

FY start	FY end	Country	Local s/n start	Local s/n end	No.	Comments
92-3915		USA	92915		1	
92-3918	92-3921	USA	92918	92921	4	
92-3923		USA	92923		1	
93-0532		USA	93532		1	
93-0534		USA	93534		1	
93-0536		USA	93536		1	
93-0538		USA	93538		1	
93-0540		USA	93540		1	
93-0542		USA	93542		1	
93-0544		USA	93544		1	
93-0546		USA	93546		1	
93-0548		USA	93548		1	
93-0550		USA	93550		1	
93-0552		USA	93552		1	
93-0554		USA	93554		1	
93-0657	93-0690	Turkey	93-0657	93-0690	34	
93-1045	93-1076	Greece	045	076	32	
94-0038	94-0049	USA	94038	94049	12	
94-0071	94-0096	Turkey	94-0071	94-0096	26	
96-0080	96-0085	USA	96080	96085	6	
97-0106	97-0111	USA	97106	97111	6	
98-0003	98-0005	USA	98003	98005	3	
99-0082		USA	99082		1	
00-0218	00-0227	USA	218	227	10	
01-7050	01-7053	USA	01050	01053	4	
02-2115	02-2122	Oman	810	818	8	
02-6030	02-6035	Chile	851	856	6	
07-1001	07-1014	Turkey	07-1001	07-1014	14	
xx-xxxx	xx-xxxx	Oman			10	

F-16D Block 50

FY start	FY end	Country	Local s/n start	Local s/n end	No.	Comments
90-0834	90-0838	USA	90834	90838	5	
90-0840	90-0849	USA	90840	90849	10	
91-0462	91-0465	USA	91462	91465	4	
91-0468	91-0469	USA	91468	91469	2	
91-0471	91-0472	USA	91471	91472	2	
91-0474		USA	91474		1	
91-0476	91-0477	USA	91476	91477	2	
91-0480	91-0481	USA	91480	91481	2	
93-0691	93-0696	Turkey	93-0691	93-0696	6	
93-1077	93-1084	Greece	077	084	8	
94-0105	94-0110	Turkey	94-0105	94-0110	6	
94-1557	94-1564	Turkey	94-1557	94-1564	8	
02-2123	02-2126	Oman	801	804	4	
02-6036	02-6039	Chile	857	860	4	
07-1015	07-1030	Turkey	07-1015	07-1030	16	
xx-xxxx	xx-xxxx	Oman			2	

F-16C Block 52

FY start	FY end	Country	Local s/n start	Local s/n end	No.	Comments
90-0809		USA	90809		1	
91-0362		USA	91362		1	
91-0370		USA	91370		1	
91-0374		USA	91374		1	
91-0386		USA	91386		1	
91-0392	91-0393	USA	91392	91393	2	
91-0401		USA	91401		1	
91-0404		USA	91404		1	
91-0413		USA	91413		1	
92-3880	92-3882	USA	92880	92882	3	
92-3885		USA	92885		1	
92-3888	92-3890	USA	92888	92890	3	
92-3896		USA	92896		1	
92-3898	92-3899	USA	92898	92899	2	
92-3902	92-3903	USA	92902	92903	2	
92-3905		USA	92905		1	
92-3908	92-3909	USA	92908	92909	2	
92-3911		USA	92911		1	

FY start	FY end	Country	Local s/n start	Local s/n end	No.	Comments
92-3914		USA	92914		1	
92-3916	92-3917	USA	92916	92917	2	
92-3922		USA	92922		1	
92-4001	92-4028	South Korea	92-001	92-028	28	
93-0531		USA	93531		1	
93-0533		USA	93533		1	
93-0535		USA	93535		1	
93-0537		USA	93537		1	
93-0539		USA	93539		1	
93-0541		USA	93541		1	
93-0543		USA	93543		1	
93-0545		USA	93545		1	
93-0547		USA	93547		1	
93-0549		USA	93549		1	
93-0551		USA	93551		1	
93-0553		USA	93553		1	
93-4049	93-4100	South Korea	93-049	93-100	52	
94-0267	94-0273	Singapore	608	614	7	Local s/n are not sequential
96-5025	96-5028	Singapore	612	615	4	Local s/n are not sequential
97-0112	97-0121	Singapore	620	646	10	Local s/n are not sequential
99-1500	99-1533	Greece	500	533	34	
01-0510	01-0524	South Korea	01-510	01-524	15	
01-8530	01-8535	Greece	534	539	6	
03-0040	03-0075	Poland	4040	4075	36	
06-0001	06-0020	Greece	001	020	20	
07-0001	07-0012	Pakistan	10901	10912	12	
08-8001	08-8016	Morocco	08-8001	08-8016	16	
10-1001	10-1016	Egypt	9751	9766	16	
xx-xxxx	xx-xxxx	Iraq			12	
xx-xxxx	xx-xxxx	Iraq			12	

F-16D Block 52

FY start	FY end	Country	Local s/n start	Local s/n end	No.	Comments
90-0839		USA	90839		1	
91-0466	91-0467	USA	91455	91467	2	
91-0470		USA	91470		1	
91-0473		USA	91473		1	
91-0475		USA	91475		1	
91-0478	91-0479	USA	91478	91479	2	
92-3924	92-3927	USA	92924	92927	4	
92-4029	92-4048	South Korea	92-029	92-048	20	
93-4101	93-4120	South Korea	93-101	93-120	20	
94-0274	94-0283	Singapore	623	691	10	Local s/n are not sequential
96-5029	96-5036	Singapore	632	694	8	Local s/n are not sequential
97-0122	97-0123	Singapore	639	640	2	Local s/n are not sequential
99-1534	99-1549	Greece	600	615	16	
99-9400	99-9451	Israel	803	898	52	Local s/n are not sequential
00-1001	00-1050	Israel	401	499	50	Local s/n are not sequential
01-0525	01-0529	South Korea	01-525	01-529	5	
01-6010	01-6029	Singapore	661	680	20	
01-8536	01-8539	Greece	616	619	4	
03-0076	03-0087	Poland	4076	4087	12	
06-2110	06-2119	Greece	021	030	10	
07-0013	07-0018	Pakistan	10801	10806	6	
08-8017	08-8024	Morocco	08-8017	08-8024	8	
10-1017	10-1020	Egypt	9821	9824	4	
xx-xxxx	xx-xxxx	Iraq			6	
xx-xxxx	xx-xxxx	Iraq			6	

F-16E Block 60

FY start	FY end	Country	Local s/n start	Local s/n end	No.	Comments
00-6001	00-6055	UAE	3026	3080	55	
00-6081		UAE	3081		1	

F-16F Block 60

FY start	FY end	Country	Local s/n start	Local s/n end	No.	Comments
00-6056	00-6080	UAE	3001	3025	25	

Appendix 2

Acronyms

A/A	Air-to-air.
AAM	Air-to-air missile.
AB	Air Base; also afterburner.
ACF	Air Combat Fighter.
ACMI	Air combat manoeuvring instrumentation.
ADC	Air Data Converter.
ADF	Air Defense Fighter.
ADG	Accessory drive gearbox.
AEF	Air Expeditionary Forces.
AEG	Air Expeditionary Group.
AESA	Active electronically-scanned array.
AEW	Air Expeditionary Wing.
AFB	Air Force Base.
AFC	Afterburner fuel control.
AFE	Alternative Fighter Engine; also aircrew flight equipment.
A/G	Air-to-ground.
AGL	Above ground level.
AGSM	Anti-g strain manoeuvre.
AGTR	Air-to-ground transition range.
AIFF	Advanced identification friend or foe.
ALICS	Avionics launcher interface computer.
AMRAAM	Advanced medium-range air-to-air missile.
AMSTEL	After-MLU Structural Enhancement of Lifetime.
AMU	Aircraft Maintenance Unit.
AMUX	Analogue multiplexer.
ANG	Air National Guard.
AoA	Angle of attack. The angle of the wing as it cuts through relative airflow.
AoS	Angle of sideslip.
ARI	Aileron rudder interconnect.
ARM	Air Refuelling Wing.
ASPIS	Advanced self-protection integrated suite.
ASPJ	Airborne self-protection jammer.
AWACS	Airborne warning and control system.
BAI	Battlefield air interdiction.
BD	Battle damage.
BFM	Basic fighter manoeuvring.
BOS	Backup oxygen supply.
BVR	Beyond visual range.
C2	Command and control.
CADC	Central air data computer.
CAP	Combat air patrol.
CAPs	Critical action procedures.
CAPES	Combat avionics programmed extension suite.
CAS	Close air support.
CCIP	Common Configuration Implementation Program.

CDU	Centre display unit.
CENC	Convergent exhaust nozzle control.
CFT	Conformal fuel tank.
CIVV	Compressor inlet variable vane.
CMS	Countermeasures switch.
CNI	Communications, navigation and IFF.
CSD	Constant-speed drive.
CUPID	Combat Upgrade Implementation Details.
DACT	Dissimilar air combat training.
DATF	Deployable Air Task Force.
DBU	Digital backup.
DEAD	Destruction of enemy air defences.
DEC	Digital electronic control.
DED	Data entry display.
DEEC	Digital electronic engine control.
DMS	Display management switch.
DTC	Data transfer cartridge.
DTE	Data transfer equipment.
ECM	Electronic countermeasures.
ECCM	Electronic counter-countermeasures.
ECS	Environmental control system.
EFS	Expeditionary Fighter Squadron.
EM	Energy-Manoeuvrability.
EOR	End-of-runway.
EOS	Emergency oxygen supply.
EPAF	European Participating Air Forces.
EPU	Emergency power unit.
EWMS	Electronic warfare management system.
FLCC	Flight control computer.
FLCS	Flight control system.
FLIR	Forward-looking infrared.
FOD	Foreign object damage.
FMS	Foreign military sales.
fpm	Feet per minute.
FPM	Flight path marker.
FS	Fighter Squadron.
FTIT	Fan turbine inlet temperature.
FW	Fighter Wing.
F-X	Fighter Unknown (*not* 'Fighter Experimental').
g	Measure of gravitational force.
GBU	Glide bomb unit.
GD	General Dynamics.
GE or **GEC**	General Electric Company.
GLOC	g-induced loss of consciousness.
GPS	Global positioning system.
HARM	High-speed anti-radiation missiles.
HOG	Hands-on gain.
HOTAS	Hands-on throttle and stick.

HSD	Horizontal situation display.		**NTISR**	Non-traditional information surveillance and reconnaissance.
HSI	Horizontal situation indicator.		**NVG**	Night vision goggles.
HTS	HARM targeting system.		**NWS**	Nose-wheel steering.
HUD	Heads-up display.		**OAF**	Operation Allied Force.
IADS	Integrated air defence system.		**OBOGS**	On-board oxygen generating system.
IAF	Israeli Air Force.		**OCU**	Operational Capability Upgrade.
IBS	Integrated broadcast system.		**ODF**	Operation Deny Flight.
ICP	Integrated control panel.		**ODS**	Operation Desert Storm.
IDM	Improved data modem.		**OEF**	Operation Enduring Freedom.
IEWS	Integrated electronic warfare system.		**OFP**	Operational flight programme.
IFE	In-flight emergency.		**OIF**	Operation Iraqi Freedom.
IFF	Identification friend or foe.		**ONA**	Operation Noble Anvil.
IFTS	Internal FLIR and target system.		**ONE**	Operation Noble Eagle.
IGV	Inlet guide vane.		**ONW**	Operation Northern Watch.
IKP	Integrated keyboard panel.		**OSW**	Operation Southern Watch.
ILS	Instrument landing system.		**PMG**	Permanent magnet generator.
INS	Inertial navigation system.		**POL**	Petroleum oil lubrication.
INU	Inertial navigation unit.		**PRI**	Primary engine control.
IOC	Initial Operational Capability.		**PSA**	Pneumatic sensor assembly.
IPE	Improved Performance Engines.		**RAM**	Radar absorbent materials.
IR	Infrared.		**RCVV**	Rear compressor variable vane.
ISAF	International Security Assistance Force.		**RDR**	Radar.
JDAM	Joint direct attack munitions.		**RFP**	Request for proposal.
JHMCS	Joint helmet-mounted cueing system.		**RLG INS**	Ring laser gyro inertial navigation system.
JFS	Jet fuel starter.		**rpm**	Revolutions per minute.
JSOW	Joint stand-off weapon.		**RWR**	Radar warning receiver; pronounced 'raw'.
kN	Kilonewtons force.		**SABCA**	Société Anonyme Belge de Constructions Aéronautiques.
kt	Knots.		**SADL**	Situational awareness data link.
kva	1,000 volt amps.		**SAM**	Surface-to-air missile.
LANTIRN	Low-altitude navigation, and targeting infra-red for night.		**SCANG**	South Carolina Air National Guard.
lb st	Pounds static thrust.		**SCU**	Systems capabilities upgrade.
LDGP	Low-drag general purpose.		**SDB**	Small diameter bomb.
LFE	Large forces employment.		**SEAD**	Suppression of enemy air defence.
LG	Landing gear.		**SEC**	Secondary engine control.
LWF	Lightweight fighter.		**SFW**	Sensor-fused weapon.
MANPADS	Man-portable air defence systems.		**SIP**	Structural Improvement Program.
MEC	Main engine control.		**SLEP**	Service Life Extension Program.
MEZ	Missile engagement zone.		**SMS**	Stores management system.
MFC	Main fuel control.		**Sqn**	Squadron.
MFD	Multi-function display.		**STAR**	Structural augmentation roadmap.
MFPG	Multinational Fighter Program Group.		**TACAN**	Tactical air navigation system.
MIL	Military power – maximum power without use of reheat/afterburner.		**TBM**	Theatre ballistic missile.
MLG	Main landing gear.		**TGP**	Target pod.
MLU	Mid-Life Update Improvement Program.		**TMS**	Target management switch.
MMC	Modular mission computer.		**TOLD**	Take-off and landing data.
MPO	Manual pitch override.		**TRs**	Training rules.
MSIP	Multi-Stage Improvement Program.		**TST**	Time sensitive tasking; also Test mode.
MSL	Mean sea level.		**UFC**	Upfront control panel.
MWS	Major weapon system.		**UHF**	Ultra high frequency.
NAS	Naval Air Station.		**USAF**	United States Air Force.
NATO	North Atlantic Treaty Organisation.		**VFR**	Visual flight rule.
NAV	Navigation.		**VHF**	Very high frequency.
NFOV	Narrow field of view.		**VID**	Visual identification.
NFZ	No-Fly Zone.		**VSV**	Variable stator vanes.
NLG	Nose landing gear.		**VVI**	Vertical velocity indicator.
nm	Nautical miles.		**WCMD**	Wind-corrected munitions dispenser.
NOTAM	Notice to airmen.		**WEZ**	Weapon engagement zone.
NSAWC	Naval Strike Air Warfare Center.			

Index